3D Television (3DTV) Technology, Systems, and Deployment

Rolling Out the Infrastructure for Next-Generation Entertainment

T0262783

OTHER AUERBACH PUBLICATIONS

AUERBACH PUBLICATIONS
www.auerbach-publications.com
To Order Call: 1-800-272-7737 • Fax: 1-800-374-3401
E-mail: orders@crcpress.com

3D Television (3DTV) Technology, Systems, and Deployment

Rolling Out the Infrastructure for Next-Generation Entertainment

Daniel Minoli

CRC Press
Taylor & Francis Group
Boca Raton   London   New York

CRC Press is an imprint of the
Taylor & Francis Group, an **informa** business
AN AUERBACH BOOK

CRC Press
Taylor & Francis Group
6000 Broken Sound Parkway NW, Suite 300
Boca Raton, FL 33487-2742

© 2011 by Taylor and Francis Group, LLC
CRC Press is an imprint of Taylor & Francis Group, an Informa business

No claim to original U.S. Government works

Printed in the United States of America on acid-free paper
10 9 8 7 6 5 4 3 2 1

International Standard Book Number: 978-1-4398-4066-5 (Paperback)

Library of Congress Cataloging-in-Publication Data

Minoli, Daniel, 1952-
    3D television (3DTV) technology, systems, and deployment : rolling out the
infrastructure for next-generation entertainment / author, Daniel Minoli.
        p. cm.
    "A CRC title."
    Includes bibliographical references and index.
    ISBN 978-1-4398-4066-5 (pbk. : alk. paper) 1. Stereoscopic television. 2. Home
theaters. I. Title.

TK6658.M558 2011
621.388--dc22

2010026073

Visit the Taylor & Francis Web site at
http://www.taylorandfrancis.com

and the CRC Press Web site at
http://www.crcpress.com

For Anna, Emma, Emile, Gabby, Gino, and Angela

Contents

Preface

Three-dimensional TV (3DTV) (also called 3D home theater) became commercially available in some markets in 2010, with an expectation for expanded penetration soon thereafter. Many vendor announcements and conferences are now dedicated to the topic. Numerous manufacturers showed 3D displays at the 2009 and 2010 Consumer Electronics Show in the United States, with new hardware expected in upcoming conferences. Many industry consortia now engage in advocacy for this technology. An increasing number of movies are being shot in 3D, and many directors such as James Cameron and Steven Spielberg embrace this upcoming trend.

This text offers an early view of the deployment and rollout of this technology to provide interested planners, researchers, and engineers with an overview of the topic. Stakeholders involved with the rollout of the infrastructure needed to support this service include video engineers, equipment manufacturers, standardization committees, broadcasters, satellite operators, Internet service providers, terrestrial telecommunications carriers, storage companies, content-development entities, design engineers, planners, college professors and students, and venture capitalists, to list a few.

This is the first practical, nonacademic book on the topic. Two other books have appeared on this topic in the recent past, but they are edited collections of fairly theoretical (signal processing, coding, etc.) and other advanced research topics. This book takes a pragmatic, practitioner's view; it is not intended for readers who need theoretical, highly mathematical, or signal-processing-oriented treatment and/or fundamental research. The focus of this text is how to actually deploy the technology. There is a significant quantity of published material in the form of papers, reports, and technical specs that form the basis for this presentation, but the information is presented here in a self-contained, organized, tutorial fashion.

Beyond the basic technological building blocks, 3DTV stakeholders need to consider a system-level view of what it will take to deploy a national infrastructure of 3DTV providers, where such providers are positioned to properly deliver a commercial-grade-quality bundle of multiple 3DTV content channels to premium-paying customers. Although there is a lot of academic interest in various subelements of the overall system, the paying public and the service providers are ultimately concerned with a system-level view of the delivery apparatus.

This text takes such a system-level view. Fundamental visual concepts supporting stereographic perception of 3DTV are reviewed. 3DTV technology and digital video principles are discussed. Elements of an end-to-end 3DTV system are covered. End-user devices are addressed. A press-time overview of the industry is provided by surveying a number of advocacy groups. Compression and transmission technologies are assessed, along with a number of technical details related to 3DTV. Standardization activities, critical to any sort of broad deployment, are identified.

The Author

Daniel Minoli has done extensive work in video engineering, design, and implementation. The results presented in this book are based on work done while at Bellcore/Telcordia, Stevens Institute of Technology, AT&T, and other engineering firms, starting in the early 1990s and continuing to the present. Some of his video work has been documented in his books, such as *IP Multicast with Applications to IPTV and Mobile DVB-H* (Wiley/IEEE Press, 2008), *Video Dialtone Technology: Digital Video over ADSL, HFC, FTTC, and ATM* (McGraw-Hill, 1995), *Distributed Multimedia through Broadband Communication Services* (co-authored; Artech House, 1994), *Digital Video* (four chapters) in *The Telecommunications Handbook* (edited by K. Terplan and P. Morreale, IEEE Press, 2000), and *Distance Learning: Technology and Applications* (Artech House, 1996).

Mr. Minoli has many years of technical, hands-on, and managerial experience in planning, designing, deploying, and operating IP/IPv6, telecom, wireless, and video networks, and data center systems and subsystems for global best-in-class carriers and financial companies. He has worked at financial firms such as AIG, Prudential Securities, and Capital One Financial, and at service provider firms such as Network Analysis Corporation, Bell Telephone Laboratories, ITT, Bell Communications Research (now Telcordia), AT&T, Leading Edge Networks, and SES Engineering, where he is director of terrestrial systems engineering (SES is the largest satellite services company in the world). At SES, in addition to other duties, Mr. Minoli has been responsible for the development and deployment of IPTV systems, terrestrial and mobile IP-based networking services, and IPv6 services over satellite links. He also played a founding role in the launching of two companies through the high-tech incubator Leading Edge Networks Inc., which he ran in the early 2000s: Global Wireless Services, a provider of secure broadband hotspot mobile Internet and hotspot VoIP services, and InfoPort Communications Group, an optical and gigabit Ethernet metropolitan carrier supporting data center/SAN/channel extension and cloud computing network access services. For several years he has been session, tutorial, and now overall technical program chair for the IEEE Enterprise Networking (ENTNET) conference; ENTNET focuses on enterprise networking requirements for financial firms and other corporate institutions.

Mr. Minoli has also written columns for *ComputerWorld*, *NetworkWorld*, and *Network Computing* (1985–2006). He has taught at New York University (Information Technology Institute), Rutgers University, and Stevens Institute of Technology (1984–2006). He was also a technology analyst at large for Gartner/DataPro (1985–2001); based on extensive hands-on work at financial firms and carriers, he tracked technologies and wrote CTO/CIO-level technical scans in the area of telephony and data systems, including topics on security, disaster recovery, network management, LANs, WANs (ATM and MPLS), wireless (LAN and public hotspot), VoIP, network design/economics, carrier networks (such as metro Ethernet and CWDM/DWDM), and e-commerce. Over the years he has advised venture capitalist firms regarding investments of $150 million in a dozen high-tech companies. He has acted as an expert witness in a successful $11 billion lawsuit regarding a VoIP-based wireless air-to-ground communication system, and has been involved as a technical expert in a number of patent infringement lawsuits (including two lawsuits on digital imaging).

1

INTRODUCTION

1.1 Overview

Just as high-definition (HD) content delivery is starting to see deployment around the world, broadcasters and bandwidth providers are being afforded the opportunity to generate new revenues with the delivery of next-generation HD entertainment with three-dimensional (3D) programming. 3D video can be visually dramatic for several genres of entertainment, including but not limited to movies and sporting events. According to recent industry surveys, a majority of people would want to watch 3D to enhance their viewing experience. 3D video has been around for over half a century in the cinematic context, and a number of 3D movies have been produced over the years; viewing of these movies has required the use of inexpensive colored or polarized "glasses." The industry is now looking to move beyond special effects in the theater and bring the technology into the mainstream of movie production and into the home. The goal of this industry is to be able to deliver 3D content over a television system and to use less obstructive (but perhaps more expensive) glasses to enhance the quality of the video experience, or eliminate the glasses altogether.

Three-dimensional TV (3DTV) (also called 3D home theater) became commercially available in a number of markets in 2010. The goal of the service is to replicate the experience achievable in 3D cinematic presentations in a more intimate home setting. A commercial 3DTV system is comprised of the following elements: capture of 3D moving scenes, scene encoding (representation), scene compression, scene transport (or media storage), and scene display. To achieve this, at least three elements are needed: (1) high-quality 3D (stereoscopic) content; (2) appropriate content delivery channels (media, satellite, or terrestrial based); and (3) high-quality home display systems. A number of manufacturers showcased a selection of 3D displays at the 2009 and 2010 Consumer Electronics Show in the United States, and newly developed hardware is expected to be introduced in upcoming trade shows. These products, including #D TV screens, 3D Blu-ray players, 3D projectors, 3D camcorders, and 3D cameras or some subset of them, may be in some consumer homes in 2010 or 2011. Vendors such as IBM, LG, Samsung, Panasonic, Sony, JVC, Mitsubishi, Vizio, Sharp and Philips are active in this space. Samsung and Mitsubishi were already shipping 3D-ready flat-panel TVs in 2008 based on digital light-processing technology from Texas Instruments.

The service was about to debut at press time. In January 2010, ESPN and Discovery Communications announced plans to launch the industry's first 3D television networks. The sports programmer was planning to introduce a 3D network in the summer of 2010, while Discovery is joining forces with Sony and IMAX for a 3D network to launch in 2011. The announcements were seen as representing a potentially game-changing addition to the TV landscape. ESPN states that its new channel—ESPN 3D—was expected to feature at least 85 live sporting events during its first year, starting in June 2010 with the first 2010 FIFA World Cup match South Africa vs. Mexico. Other 3D events were expected to include up to 25 World Cup matches, the 2011 BCS National Championship Game, college basketball and football, and the Summer X Games. "ESPN's commitment to 3D is a win for fans and our business partners. ESPN 3D marries great content with new technology to enhance the fan's viewing experience and puts ESPN at the forefront of the next big advance for TV viewing," stated ESPN representatives during the announcement. ESPN had been testing 3D for more than two years prior to this announcement [SZA201001]. The Master Golf Tournament 2010 was made available in 3D on ESPN, CBS, Cox, and the BBC. Verizon FiOS was planning to offer 3D programming by the end of 2010. BSkyB announced it was starting stereo 3D broadcasts in the United Kingdom in 2010, British Channel 4 was planning to offer a selection of 3D programming at the end of 2009, and other broadcasters and satellite operators are planning to do the same. Sony announced plan to release highlights from the 2010 World Cup on Blu-ray disc (BD); the Blu-ray release of *Titanic*, *Terminator*, and *Avatar* are also expected in 2010. The ability to use polarized or shutter glasses, rather than the traditional red and green glasses, is expected to represent a positive driver to the introduction of the technology in the home. However, the need to use an accessory at all may be somewhat of a hindrance for some; systems that do not require accessories are slightly behind in the development cycle compared with glasses-based systems. A subtending goal of the 3DTV rollout initiative is to provide quality two-dimensional (2D) performance to existing customers accessing but not exploiting the 3D content.

In recent years there has been high growth in the area of 3D cinema: In 2006, 330 3D screens were installed in the United States, in 2007 this number increased to 768, and in 2008 it doubled. By 2011 more than 4,300 3D screens are expected [3DM200901]. Worldwide, the number of 3D cinema screens was more than 1,300 at the end of 2007 and was predicted to grow to over 6,000 by 2009 [BAR200901]. The Consumer Electronics Association (CEA) estimates that about 2.2 million 3D TV sets will be sold in 2010, and that more than 25 percent of sets sold in 2013 will be 3D enabled. Early adopters in the United States make up about 5 million households that could adopt 3D TV within three years; after that, about 20 million additional homes could sign up for 3D "pretty quickly," before the technology then goes mass market in about five to ten years [SZA201001].

A number of Hollywood studios have announced major productions in 3D and an increasing number of movies are now being shot in 3D; *all* future Pixar movies are reportedly being shot in 3D and many directors (including James Cameron and Steven Spielberg) are reportedly embracing this trend [SOB200901]. Hollywood studios had several new 3D titles ready in the summer of 2010. The Blu-ray Disc Association (BDA) announced in late 2009 specifications for creating full 1080p 3D Blu-ray disc content, and the first 3D-enabled Blu-ray players made its debut at the Consumer Electronics Show in January 2010.

A joint 2008 consumer study of the CEA and the Entertainment and Technology Center at the University of Southern California concluded that 3D technology is now positioned "to become a major force in future in-home entertainment." According to the study, an estimated 41 million U.S. adults have seen a 3D movie in theaters in the past 12 months, and nearly 40 percent (of the consumers polled) said they would prefer to watch a movie in 3D than watching that same movie in 2D [TAR200901].

There is now a renewed industry interest in the 3DTV research primarily due to the advances in low-cost 3D display technologies. Propitiously, there is also interest in this topic by the standards-making organizations. The Moving Picture Experts Group (MPEG) of the International Organization for Standardization/International Electrotechnical Commission (ISO/IEC) is working on coding formats for 3D video and has already completed some of them. As far back as 2003, the 3D Consortium (3DC) with 70 partner organizations was formed in Japan and more recently, a number of new activities have been started, including the 3D@Home Consortium, the Society of Motion Picture and Television Engineers (SMPTE) 3D Home Entertainment Task Force, the Rapporteur Group on 3DTV of ITU-R Study Group 6, and the TM-3D-SM group of the Digital Video Broadcast Project (DVB). At face value, it would appear at press time that the European community is ahead of the rest of the world at the system-research/system-design level, while the Japanese/Korean manufacturers are taking the lead at the TV display-production level, and from a commercial service delivery the United States appears to be ahead.

Stereo means "having depth, or three dimensions"; stereo vision is the process whereby two eye views are combined in the brain to create the visual perception of a single three-dimensional image. There are two commercial-grade approaches to 3DTV: the use of *stereoscopic* TV displays, which requires special glasses to watch 3D movies, and the use of *autostereoscopic* TV displays, which show 3D images in such a manner that the user can enjoy the viewing experience without special accessories [ONU200701]. Short-term commercial 3DTV deployments, and the focus of this book, are on stereoscopic 3D (S3D) imaging and movie technology. (Autostereoscopic technology may be appropriate for mobile 3D phones and there are several initiatives to explore these applications and this 3D phone-display technology.) S3D (also known as stereoscopy or stereoscopic imagery) uses the characteristics of human binocular vision to create the illusion of depth, making objects appear to be in front of or behind the screen. The technique relies on presenting the right and left eyes with two slightly different images which the brain automatically blends into a single view. Subtle right–left dissimilarities in the images create the perception of depth and can be manipulated to creative advantage [AUT200801]. It should be noted up front that holographic* 3DTV with high-quality optical replica of artifacts, objects, and people appearing to float in space or standing on a tabletop-like display, where viewers are able to look or walk around the images to see them from different angles, is not within practical technical, content mastering, or commercial reach at this time for home TV reception, although there are proponents. Besides applications in cinema (and emerging TV usage), 3D viewing technology has been used for a number of years by players of video games by utilizing head-mounted displays that provide a fully immersive experience.

The basic technical challenges to provide a routinely available, commercial 3DTV service relate to the reliable/inexpensive capture, representation (encoding, compression), transmission, display, synchronization, and storage of elementary streams (ES) that comprise the content (ESs contain audio information, video information, or data—such as closed caption electronic program guides—that is formatted into a packetized stream).

The physiological reason for the human sense of depth has been understood for over 100 years. Based on that understanding, a number of techniques for recreating depth for the viewer of photographic or video content have been developed. Fundamentally the technique known as "stereoscopy" has been advanced, where two pictures or scenes are shot, one for each eye, and each eye is presented with its proper picture or scene, in one fashion or another. We note here that neither 3D stereoscopic viewing nor 3DTV is "brand new" per se: Stereoscopic 3D viewing techniques are almost as old as their 2D counterparts. Soon after the invention of movies in the 1900s, stereoscopic images were periodically shown at public theaters. The world's first stereoscopic motion picture and camera were made by William Friese-Green in 1893 [STA200801]. The means to achieve stereoscopic display has migrated over the years from anaglyth to polarization; anaglyth is a basic and inexpensive method of 3D transmission that relies on inexpensive colored glasses, but its drawback is the relatively

---

* The concept of a hologram is built on the pattern of light formed by interference between two crossing light rays: a "signal" (or "object") beam that restores information that was recorded inside photosensitive storage media, and a "reference" beam. The three-dimensional picture formed by this light configuration is called a hologram [HTH200701].

low quality. This technical progression eventually led to the routine display of 3D in theaters, where stereoscopic movies now are shown fairly regularly. In Europe and Japan anaglyphic TV broadcasts (for example, with two-channel, phase alternate line [PAL] demonstrations) were trialed in 1983, 1985, and 1987). To mention a few more recent initiatives, an experimental stereoscopic 3D-HDTV system was documented in 1999 by the NHK* Science & Technical Research Laboratories (NHK-STRL). Also, there was a Korean broadcasting experiment in 3D-HDTV with the broadcast of 2002 FIFA World Cup. In 2009 there was a satellite transmission in Europe of a live music performance that was captured, transmitted, and projected on prototype models of stereoscopic 3DTVs, and there was also a satellite transmission of a live opera performance from a French opera house to electronic cinema sites throughout France [BAT200901].

A considerable amount of research has taken place during the past 30-plus years on 3D graphics and imaging;† most of the research has focused on photographic techniques, computer graphics, 3D movies, and holography. More recently and with the advent of HDTV, research is starting to emerge on the use of HDTV techniques to deliver stereographic content directly to the home. With regard to actual commercial deployment of home 3DTV services at this time or in the near future, a key development has been the relatively recent introduction of HD services that provide a more realistic experience by increasing spatial resolution and widening the viewing angle. 3DTV focuses on a sensation of depth; psychological studies have been undertaken in the past couple of decades, validating that a wide screen is necessary to achieve adequate 3D experiences—hence the genesis of stereoscopic HDTV services, that is to say 3DTV as we know it today.

---

* NHK stands for Nippon Hoso Kyokai, which translates to Japan Broadcasting Corporation.

† The field of imaging, including 3D imaging, relates more to the static or quasi-static capture/representation (encoding, compression)/transmission/display/storage of content)—for example, photographs, medical images, CAD/CAM drawings—especially for high-resolution applications. This topic is not covered here. The interested reader may consult, among other texts, D. Minoli, Imaging in Corporate Environments, Technology, and Communications, McGraw-Hill, 1994.

The two video views required for 3DTV can be compressed using standard video compression techniques in what is called a simulcast transmission approach to 3DTV. However, more effective compression is sought. Moving Picture Expert Group 2 (MPEG-2) encoding is widely used in digital TV applications today and H.264/MPEG-4 Advanced Video Coding (AVC) is expected to be the leading video technology standard for digital video in the near future. Extensions have been developed recently to H.264/MPEG-4 AVC and other related standards to support 3DTV; other standardization work is underway. The compression gains and quality of 3DTV will vary depending on the video coding standard used. While inter-view prediction will likely improve the compression efficiency compared to simulcasting (transmitting the two views), new approaches, such as asymmetric view coding, are necessary to reduce bandwidth requirements for 3DTV [CHR200601].

Clearly 3DTV will require more bandwidth of regular programming, perhaps even twice the bandwidth; some newer schemes such as "video+depth" may require only 25 percent more bandwidth compared to 2D, but these schemes are not the leading candidate technologies for actual deployment in the next two to three years. Techniques now being advanced involve side-by-side juxtaposition of left/right frames where the two images are reformatted and compressed (at the price of reduced resolution) to fit into a standard (HD) channel. If HDTV programming is broadcast at high quality—say, 12–15 Mbps using MPEG-4 encoding—3DTV will require 24–30 Mbps when the simulcast approach is used.* This data rate does not fit a standard over-the-air digital TV (DTV) channel of 19.2 Mbps, and will also be a challenge for non-fiber-to-the-home (non-FTTH) broadband Internet connections. However, one expects to see the emergence of bandwidth reduction techniques, as alluded to above. On the other hand, direct-to-home (DTH) satellite providers, terrestrial fiber-optic providers, and some cable TV firms should have adequate bandwidth to support the service. For example, the use of digital video broadcast, satellite second generation (DVB-S2) allows a transponder to carry 75

---

* Some HDTV content may be delivered at lower rates by same operators, say 8 Mbps; this rate, however, may not be adequate for sporting HDTV channels and may be marginal for 3D TV at 1080p/60Hz per eye.

Mbps of content with modulation using an 8-point constellation and twice that much with a 16-point constellation.* However, the tradeoff would be (if we use the raw HD bandwidth just described as a point of reference) that a DVB-S2 transponder that would otherwise carry 25 channels of standard definition video or six to eight channels of HD video would now only carry two to three 3DTV channels. For this reason, the simulcast approach is not being seriously considered at this time; the reduced resolution side-by-side approach is preferred in the short term.

Consumer electronics original equipment manufacturers (OEMs), content owners, service providers, retailers, and consumers are the stakeholders in this new 3DTV service deployment. As is the case with any new technology, industry economics and business models play a key role in the early stage of deployment. There are also many technical and design questions that need to be answered before widespread 3DTV deployment can take place; these are related to postproduction 3D mastering, delivery options, and home TV screens. Mastering of images for 3DTV will likely require 1920x1080 pixel resolution at 60 frames per second and per eye. At this juncture neither the mastering technology nor the end-to-end transmission systems are standardized. Several manufactures are aiming at the 3DTV market, but each has advocated, so far, a different approach. As a minimum there has to be standardization for image formatting of the source materials; content can then be delivered by content distributors over a number of (or all) distribution channels: from physical media to terrestrial, satellite, cable, and other streaming services. A milestone was reached in 2009, in the view of some, when SMPTE defined the basic requirements for a stereoscopic 3D Home Master standard. In general, there is interest in developing a backwards-compatible and flexible broadcast 3DTV system.

Figure 1.1 depicts a basic 3DTV system. As can be inferred from the figure, an end-to-end upgrade of the current environment is

---

* The Digital Video Broadcasting Project (DVB) is an industry-led consortium of over 250 broadcasters, manufacturers, network operators, software developers, regulatory bodies, and others in over 35 countries committed to designing open technical standards for the global delivery of digital television and data services. Services using DVB standards are available on every continent with more than 500 million DVB receivers deployed.

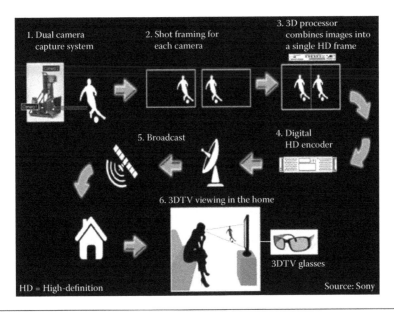

Figure 1.1 Basic 3DTV system.

needed in order to be able to support the 3DTV service. This includes image (content) capture, signal processing and encoding, transmission, and screen rendering. A 3DTV signal can be seen as a container that carries two content streams (two TV programs): one for the right eye and one for the left eye.* For full resolution acquisition where each stream is captured at 24 frames per second (fps), this clearly means that each second of content will have 48 frames. For 60p (progressive) HD acquisition, each signal will be captured at 60 fps, generating a combined stream of 120 fps. This type of stereogram allows images to be printed, projected, or viewed on a TV or computer screen to provide 3D perception; the two images are superimposed but are separated, such that each eye sees only the desired image. An early approach to 3D is via an anaglyth, namely, with the use of colored filters and viewing spectacles (commonly red and cyan, or red and green—to the naked eye, the encoded 3D image looks overlapping, doubled, and blurry); usually, the image for the left eye is printed in red ink and the right-eye image is printed in green ink [3DA201001]. Note that this early anaglyph technique produces an experience that is inferior to

---

* Another approach is to carry some data (say, 2D content for one eye) plus a channel of metadata to generate the 3D signal.

the cinematic 3D experience in a theater. Most of the current 3DTV displays for televisions (and cinema) require wearing either polarized or shuttered glasses. TV displays likely will need about $100 in new silicon to support 3DTV. Products offering autostereoscopic imaging without the need of special glasses are emerging; for example, a line of 3D televisions initially planned to be brought to market by Philips used very small lenses placed over the red, blue, and green pixel points that make up the television screen. The lenses cause each individual pixel to project light at one of a series of nine angles projecting from the display [ISS199901].

As noted, technical advances—or at the very least, system upgrades—are required along the entire chain of Figure 1.1. To illustrate this point, advances are being made for content capture; for example, Sony recently introduced a single-lens camera that is able to capture 3D images. The majority of existing 3D setups use two-camera systems to record images tailored specifically for the left and right eyes of the viewer. The new Sony camera takes a single image that is split by mirrors and recorded on two sensors, resulting in a "smoother" picture. With this program acquisition, viewers are able to watch the 3D images using special polarized glasses; without these glasses, they will just see normal 2D television. End-to-end interoperability, from content creation to the home user, is critical to the success of this nascent service. Furthermore, quality of the delivered video will depend upon appropriate quality support of each element in the chain: if any one link in the chain is deficient in some fashion, then the overall quality will suffer accordingly. In the short term, 3D content of choice may be available at home from a Blu-ray disc player or a PC file; in the longer term, the content will be available at the set-top box (STB), this being fed from a terrestrial cable TV system, a satellite DTH link, a broadband Internet connection, or, eventually, also from terrestrial over-the-air broadcasting.

To generate quality 3D content, the creator needs to control the *depth* and *parallax* of the scene, among other parameters. Depth perception is the ability to see in 3D to allow the viewer to judge the relative distances of objects; depth range is a term that applies to stereoscopic images created with cameras. Parallax is the apparent change in the position of an object when viewed from different points—namely, the visual differences in a scene when viewed from different points. A 3D display (screen) needs to generate some sort of parallax that, in turn, creates a stereoscopic sense. There are a number of ways to create 3D content, including (1) computer-generated imagery (CGI), (2) stereo cameras, and (3) 2D-to-3D conversions. CGI techniques are currently the most technically advanced, with well-developed methodologies and tools to create movies, games, and other graphical applications, and the majority of cinematic 3D content is comprised of animated movies created with CGI. Camera-based 3D is more challenging. A two-camera approach is the norm at this time; another approach is to use a 2D camera in conjunction with a depth-mapping system. With the two-camera approach, the two cameras are assembled with the same spatial separation to mimic how the eye may perceive a scene. The technical issues relate to focus/focal length, specifically keeping in mind that these have to be matched precisely to avoid differences in vertical and horizontal alignment and/or rotational differences (lens calibration and motion control must be added to the camera lenses). Conversion of 2D material is the least desirable but perhaps it is the approach that could generate the largest amount of content in the short term. Some note that it is "easy to create 3D content, but it is hard to create good 3D content" [CHI200901].

From a content-creation/editing perspective, first-iteration approaches of having "stuff fly into the face of the viewer" are now being replaced with early-stage "viewer in the scene" immersion content that might be classified as being closer to virtual reality (VR). Future developments may include the creation of rich, continuous, multi-individual virtual environments, interactive immersive animation environments, immersive VR,* telepresence, mixed reality (MR)/augmented reality (AR), and other synthetic sensory environments that as a group we call real virtuality (RV), but the commercialization of these advances may be a decade off or more.

However, the reader and/or viewer should not mistake (or compare) CGI-created "viewer in the scene" material, popular in cinematic 3D movies, with simple 3D perception: The former is very engaging, compelling, and intensive while the latter basically provides a more

---

* Telepresence refers to the process of interacting with a remote real world while immersive virtual reality refers to the process of interacting with a digital world [ALR200901].

natural and realistic view of the content. Seeing a soccer or golf game in 3D will not be as dramatic as a scene in the Avatar movie—other sporting events such as wrestling or boxing may perhaps provide a more vivid 3D experience if the cameras are right in the ring.

1.2 Background and Opportunities

There are ardent proponents of the technology and its commercial opportunity, while others take a more cautious view. This section provides an array of industry snapshots to give the reader a sense of where the service and technology stands.

The technology is being presented this way by proponents:

[T]he next killer application for the home entertainment industry—3DTV … will drive new revenue opportunities for content creators and distributors by enabling 3D feature films and other programming to be played on their home television and computer displays—regardless of delivery channels. [SMP200901]

Others offer observations such as these:

The recent popularity of 3D movies in North America has fueled the global 3D boom, and consortiums related to 3D including 3D@Home in North America, 3D4YOU in Europe, 3DFIC in Korea and C3D in China were established and commitment to lead the 3D industry started. [3DC200901]

3DTV augments the traditional TV technology by showing the viewer not only sequences of 2D images but streams of three-dimensional scene representations. To the viewer at home this will mean a completely new media experience. He will perceive the displayed events in a more immersive way, and he may even get the chance to choose his own viewpoint to watch the displayed events. In the future, 3D movies will become a standard and provide enhanced interactivity options, e.g., by allowing the user to navigate through the scenes. [3DT200701]

Throughout [2009], moviegoers have shown an overwhelming preference for 3D when presented with the option to see a theatrical release in either 3D or 2D. We believe this demand for 3D content will carry over into the home now that we have, in Blu-ray Disc, a medium that can deliver a quality Full HD 3D experience to the living room. ... In 2009 we saw Blu-ray firmly establish itself as the most rapidly adopted packaged media format ever introduced. We think the broad and rapid acceptance Blu-ray disc already enjoys with consumers will be a factor in accelerating the uptake of 3D in the home. In the meantime, existing players and libraries can continue to be fully enjoyed as consumers consider extending into 3D home entertainment. [SHI200901]

The recent box office success of "Avatar," which recently passed $1 billion worldwide and is set to become the #2 movie of all time behind director James Cameron's own "Titanic," has helped prove the 3D format can draw a stunning number of viewers. [SZA201001]

As early as in the 1920s, John Logie Baird, one of the TV pioneers, dreamed of developing high-quality, three-dimensional (3D) color TV, as only such a system would provide the most natural viewing experience. Today, eighty years later, the first black-and-white television prototypes have evolved into high-definition digital color TV, but the hurdle of 3D still remains to be overtaken. New hope arises from recent advances in a number of key technologies, with the following developments being of particular importance: (a) the introduction and increasing propagation of digital TV in Europe, Asia and the United States; (b) the latest achievements in the area of single- and multiview autostereoscopic 3D display technologies; (c) the increased interest in the investigation of the human-factors requirements for high-quality 3DTV systems. [Author's note: Many of these issues are still valid.] [FEH200401]

3DTV is one of the "hottest" subjects today in broadcasting. The combination of the audience's "wow" factor and the potential to launch completely new services makes it an attractive subject for both consumer and professional. There have already been broadcasts of a conventional display-compatible system, and the first HDTV channel compatible broadcasts are scheduled to start in Europe in the Spring of 2010. [DVB201001]

Hollywood studios are eager to find new ways to gain revenues from an increasing number of 3D titles they are developing for the cinema. Engineers have explored ideas for 3D on television for years, and 3DTV demos have long been a staple of major exhibitions for consumer electronics giants. But the latest moves indicate big industry organizations may think the time is right to plough the road for tomorrow's mainstream products. [MER200801]

Of all the new High Definition Multimedia Interface (HDMI) Version 1.4 features, 3D is getting the most interest from the broadcasters. … Everybody is lined up behind 3D, so it will be a big launch next year [2010]. [MER200901]

3D media is clearly the wave of the future! [ANA200901]

3D multimedia applications are receiving increasing attention from researchers in academia and in industry. This is in part due to new developments in display technologies, which are providing high quality immersive experiences for prices within the range of consumers. 3D video has become a strong candidate for the next upgrade to multimedia applications, following the introduction of HD video. [HEW200901]

But others offer observations such as these:

3DTV is coming to a living room near you. But will the technology spur a consumer spending spree like digital and high-definition TV did before it? Or will 3D end up being the next big flop? One thing is clear: TV manufacturers need something new to get people buying TVs. [REA200901]

In a wide range of demos, companies will claim at the Consumer Electronics Show in January 2010 that stereoscopic 3D is ready for the home. In fact, engineers face plenty of work hammering out the standards and silicon for 3DTV products, most of which will ship for the holiday 2010 season. [MER200901]

At a recent (2009) CEA Industry Forum, the focus has been on consumer electronics retail trends (such as changes in channel dynamics), 3DTV technology, green technology, and social media. CEA takes the tentative position that:

3DTV technology is demonstrating clear success at movie theatres and will gradually evolve into other facets of consumers' viewing habits; 3D TV is similar to HDTV in that consumers are more likely to want it once they have truly experienced it. But the industry needs to have reasonable expectations for 3DTV. It is gaining momentum but may not hit critical mass for several years. [CEA200901]

The ITU noted recently that:

It has proven somewhat difficult to create a 3D system that does not cause "eye fatigue" after a certain time. Most current-generation higher resolution systems also need special eyeglasses, which can be inconvenient. Apart from eye-fatigue, systems developed so far can also have limitations such as constrained viewing positions. Multiple viewpoint television systems are intended to alleviate this. Stereoscopic systems also allow only limited "production grammar." … One should not under-estimate the difficulty, or the imagination and creativity required, to create a near "ideal" 3DTV system that the public could enjoy in a relaxed way, and for a long period of time. [ITU200801]

Creating a 3D television system that can be viewed in comfort poses great challenges. We can—and should—examine "stereoscopic systems" to see how well we can make them work, and how we can arrange compatibility with normal television channels. It may be possible to achieve compatible 3D television systems which can be watched comfortably for the length of a program. But at the same time, we need to continue the fundamental research into [holography] to make possible 3D television which really is equivalent to "being there." [DOS200801]

Practitioners also note that:

The production pipeline for 2D television has developed into a mature and well-understood process over many years. Scenes are recorded with cameras from single-view points, captured image streams are post-processed, transferred to receivers, and displayed on planar screens. In contrast, the production process for 3D television requires a fundamental rethinking of the underlying technology. Scenes have to be recorded with multiple imaging devices that may be augmented with additional sensor technology to capture the three-dimensional nature of real scenes. In addition, the data format used in 3D television is a lot more complex. Rather than normal video streams, time-varying computational models of the recorded scenes are required that comprise of descriptions of the scenes' shape, motion, and multiview appearance. The reconstruction of these models from the multiview sensor data is one of the major challenges that we face today. Finally, the captured scene descriptions have to be shown to the viewer in three-dimensions which requires completely new display technology. [ROS200801]

When the 3DTV goal is stated as follows, it is clear that this is an unattainable milestone at least for the next decade or so. Something more pragmatic is in order, as described in this text, at least for the next two to five years, to mid-decade:

The ultimate goal of the viewing experience is to create the illusion of a real environment in its absence. If this goal is fully achieved, there is no way for an observer to distinguish whether or not what he sees is real or an optical illusion. [ONU200601]

Market observers state that 3DTV will be available commercially relatively soon in a number of key regions of the world, although it is expected to gradually grow only to about 30 percent penetration over time. According to GigaOM Pro, a U.S. consumer technology research group, the 3DTV market will see shipments of nearly 50 million 3D televisions by the year 2013. The firm predicts the consumer demand will be so great that media and content providers will rush to reformat old material to make it 3D ready in order to satisfy the demand. Research firm DisplaySearch forecasted the 3D-display/3DTV market to reach $1.1 billion in 2010 and grow to $15.8 billion by 2015 [SOE200901]. The reader should note, however, that market forecasts are notoriously unreliable.*

The question, however, remains if the consumers are willing to make new investments after the flat-panel TV expenditures of the last decade. The macroeconomics issues of the time may impact 3DTV development. For example, Philips discontinued work on the WOWvx line in 2009. The WOWvx website stated:

Most of you are probably aware by now that Philips has decided to shut down its 3D operation due to current market developments. As a result, WOWvx.com will be closing its doors and go offline at the close of business (5pm CET) on June 30th, 2009. This is a difficult time for many of us and we realize that it is not any easier for a lot of our community.

As noted earlier, the 3D market was pioneered by the film industry. 3D photography, 3D cinema, and 3DTV actually have a long history as concepts; in fact, stereoscopic 3D versions of these common visual media are almost as old as their 2D counterparts. For example, stereoscopic 3D photography was invented as early as 1839 and the first examples of 3D cinema (short movies) were available in the early

---

* Based on years of experience, this author has developed the following heuristics: Take a forecast for x revenue/units/subscribers y years out; well, a safe bankable bet is for x/2 revenue/units/subscribers 2y years out.

1900s. Early 2D television concepts were developed in the 1920s and by 1929 stereoscopic 3DTV concepts were demonstrated. However, while the 2D versions of photography, cinema, and TV have flourished commercially and technologically, their 3D counterparts have not been commercially deployed on a broad scale [ONU200801]. In 1953 *Man in the Dark* was the first commercial full-length 3D movie;* however, early technology caused unsteady images that induced nausea. A second attempt at commercial introduction of 3D was made in the 1970s using stereoscopic images that required users to wear red and green glasses. Although the image was steady there was considerable loss of color quality and, as a result, the service failed yet again to take off. The latest round (called the "3D Wave" by some) kicked off in 2003 with the release of the film *Ghosts of the Abyss*. During the past few years about 30 theatrical 3D movies have been released. A short partial list includes *Up, G-Force, Coraline, Journey to the Center of the Earth, Hannah Montana, Jonas Brothers, Fly Me to the Moon, Bolt, Ice Age, Dawn of the Dinosaurs, Toy Story 1* and *2*, and *Avatar* (a game based on this movie was released in December 2009).

As observed above, there is now an industry expectation of technology pull-through from cinema to home. The 3D movie experience is seen by proponents as driving awareness, interest, and desire to have this content available on home theater systems. It should be noted, however, that systems that cause any discomfort or restrain the viewer will likely not be broadly accepted. Meanwhile, the technology to capture in 3D—or to create a virtual 3D image using conventional cameras—has also been getting less expensive and is now affordable by some traditional TV and satellite broadcasters [BBC200901]. Content creation tools are now becoming available that might spur a new phase of 3D content creation. This may be similar to the video editing tools that have emerged in the past 10 to 15 years that have already revolutionized TV as we knew it. Furthermore, until recently,

---

* *Man in the Dark* opened at the Globe Theater in New York City on April 8, 1953. It was the first 3D movie released by a major Hollywood studio, beating The House of Wax to the screen by two days. It took Hollywood a while to embrace 3D technology; the earliest known 3D feature film not made by a mainline studio was *The Power of Love* in 1922. But 3D had already been around for a long time. Charles Wheatstone made the world's first stereoscopic viewer in 1838, basing his invention on theories of perspective dating to the Renaissance [LON200801].

Figure 1.2   3DTV as represented in planar brochures of technology providers. (Courtesy: Philips. http//www.inition.co.uk/ignition/product.php?URL_=product_stereovis_philips. With permission.)

3D cameras and other equipment associated with 3D production were available only from boutique production firms, but now they are starting to become available in rental house inventories [JOH200901]. Figures 1.2 and 1.3 depict illustrative examples of how technology providers are representing 3DTV in their marketing brochures.

Forces pulling 3D into the home include the following:

- "Hollywood goes 3D" and broader 3D content generation— Hollywood studios are driving 3D to the home as a way to make more money on a growing number of successful 3D titles at the theater [MER200901]
- Digital television (DTV) broadcast transition
- Introduction of HD services—the maturing of high-resolution video techniques have made 3D production economically feasible
- Consumer desire for improved TV and purchases of flat-panel TV sets
- Broadband Internet access
- Digital cinema
- Need for new revenue by service providers

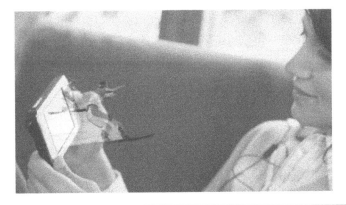

Figure 1.3 Other 3DTV as represented in planar brochures of technology providers. (Courtesy: Nokia. http://research.nokia.com/research/mobile3D. With permission.)

Retardant factors and technological deficiencies of 3DTV currently include the lack of practical, fully electronic means of 3D scene capture, and relatively inexpensive 3D scene display units, preferably without requiring accessories such as viewing glasses. In a 3D cinema environment, wearing glasses is more acceptable to viewers, since this is only for a limited amount of time and theaters typically provide excellent visual quality. Wearing glasses at home, especially for several hours at a time, may be problematic. Even people who wear vision glasses might not regularly use them at home. Furthermore, the viewer's habits and requirements in the home are very different from cinema applications, where everyone is sitting in a comfortable chair for a given time and paying full attention to the presented content without moving about or interacting with each other [3DP200802]. This observation may hold some pragmatic truth, although many would argue to the contrary:

> Until someone comes up with a way to put hundreds of millions of new 3D TVs in homes which can use polarized glasses or LCD [liquid crystal display] shutter glasses, anaglyph is going to be the only way for mass distribution of full color high quality 3D over cable, satellite, the web or on DVD. [STA200901]

Revenue opportunities will ultimately be a key driver for the widespread introduction of the technology. 3D movies have commanded a premium at theaters. It should be noted that about half the revenue of the movie producers ("Hollywood") is derived from sources other than theaters, such as DVDs and Blu-ray discs; this implies that content developers may be interested in tapping into that market opportunity to bring 3D content to the home.

To support these opportunities, note that a specification for 3D content on Blu-ray discs was finalized at the end of 2009. The specification allows for encoding of stereoscopic video content in Blu-ray discs; with an overhead of about 50 percent compared with the status quo, the video stream will be able to store both right-eye and left-eye images with a reasonably good resolution. The new 3D-encoded Blu-ray discs will be compatible with current Blu-ray players, and the 3D players will be compatible with older 2D discs. The specification also allows for "enhanced graphic features" in Blu-ray systems that will allow for the creation of immersive 3D menus for navigating the Blu-ray disc's content, and support for 3D subtitles positioned in 3D within a scene [SOB200901]. Also, the high-definition multimedia interface (HDMI) licensing group recently promulgated the HDMI Version 1.4 spec that supports the top/bottom format (discussed in Chapter 4) that many broadcasters are expected to use. The HDMI 1.4 format overlays information about left- and right-eye images onto a single frame to save bandwidth, but with some loss of resolution. The HDMI 1.4 spec does not list a mandatory format for broadcasters although broadcasters seem to be in favor of a top/bottom implementation. An HDMI implementation of the top/bottom format will be defined in a meeting of the group in early 2010.*

---

* Carriers are looking at techniques such over/under and side-by-side to reduce transmission bandwidth from 100% overhead in simulcasting, to 50% overhead for the BD encoding, to 25% overhead in video+depth approaches, to 0% overhead with these just-named approaches. The structure of the top/bottom 3D frame-compatible format (also referred to as over/under) is made of two images that share the same video frame. In order to achieve that, the resolution of the original images is reduced to half of the video lines on the vertical resolution axis (540 lines × 1920 pixels image for the left eye plus 540 × 1920 image for the right eye = 1080 × 1920 of the typical video frame for HD). In other words, half of the quality of the original pair of 3D images is discarded even before is compressed (with MPEG-2, MPEG-4, etc.) for distribution by satellite and cable (and planned by terrestrial broadcast). If another 3D frame-compatible structure is used by the service provider, such as the side-by-side, the video frame is still made of two half-resolution images, but split over the horizontal resolution axis, having 1080 lines × 960 pixels each, both totaling 1080 × 1920 [MAE201001].

Although the 3D technology is far from being completely specified or being functionally "perfect," strides have been made recently to improve the viewing experience, especially for the theater where stereo technologies, such as RealD 3D (discussed in Chapter 3), are considered to be of good quality. The goal of the proponents and the content/equipment suppliers is to translate that experience to the home. There are a number of technical challenges to create the formats, the encoding, and distribution of 3D content. Today, 3DTV is still in its maturation phase. Many technological and computational problems in scene acquisition, scene reconstruction, and scene display are either unresolved or bring today's technology to its limits. Furthermore, the problems to be solved require expertise from many different areas in science and engineering, ranging from computer science and physics to electrical engineering [3DT200701]. Some of the current areas of industry activities include international standardization of 3D, 3D displays, 3D content production, and development of 3D safety guidelines.

As we have already discussed, availability of content will be critical to the introduction of the 3DTV service (whether this is initially via broadcast or via optical media). It should be noted that 3D content is more demanding in terms of production, not only at the technical level but also at the artistic level. Filming techniques are generally more difficult in 3D. This applies not only to pre-edited material (such as movies), but even more so with real-time sporting events (note that real-time 3D video communication is also the most challenging application scenario for 3D video processing algorithms). For example, practitioners make the observation that fast cuts may be difficult, scenes with long focal lengths require more technical thought, and capture of depth should be given due consideration (for example, slowly changing the depth from one scene to the next will minimize possible eyestrain) [3DA201001]. Additionally, one needs to take into consideration the ultimate size of the screen where the content will be viewed, although this is not altogether unique to 3D. If the content is seen in a movie theater, the 3D volume is large and will not cause eye strain; but if the same content is viewed in a small TV screen, the separation between the two visual tracks may be excessive in relative terms and eye strain may result. Consequently, material created for cinema screens may have to be scaled down for TV screens. Conversely, as

Table 1.1   Press-time Players (Partial List)

3ality Digital
DDD
Dolby Laboratories
Fraunhofer Institute for Telecommunications—Heinrich-Hertz-Institut
HDLogix
LG
In-Three
Industrial Technology Research Institute (ITRI)
Nvidia
Panasonic
Sonic Solutions
Sony
Texas Instruments
THX

may be obvious, material that has been developed for a TV screen may not be appealing on a large cinema screen. It is easier to scale from big screen to small screen than conversely. Although these issues require due attention for non-real-time programming, the situation is even more demanding for real-time events, such as sporting events; there have been a number of 3D pay-per-view real-time events with various degrees of ultimate user-experience quality.

3D research has been undertaken for many years in the United States, France, the United Kingdom, Italy, Canada, Japan, Germany, and Russia. Table 1.1 provides a partial list of 3DTV players as of press time. LG announced at the end of 2009 the world's first mass-produced, full-HD, 3D-capable monitor (a 23-inch monitor measured diagonally). Korea reportedly plans to introduce full-HD 3D broadcasts; satellite-based 3D broadcasting trials are set to take place in the United Kingdom and Japan. LG reportedly expects to sell 3.8 million 3D LCDs by 2011, and the Korean manufacturer forecasted sales of 400,000 3D units in 2010 and 3.4 million in 2011—the company has announced it will introduce a "wide range" of 3DTV sets measuring between 42 and 72 inches diagonally [SAV200901]. Figure 1.4 depicts an illustrative example of a 3DTV monitor. LG was partnering with Korean broadcaster SkyLife (for 3D content creation) and with DTH satellite providers. The game makers have been delivering 3D games for a number of years; hence while there may be a dearth of 3D video product to watch at this time, a game machine

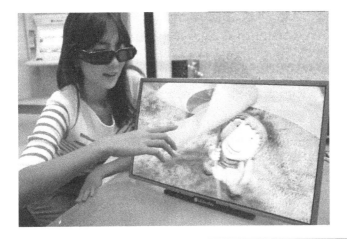

Figure 1.4   LG 3DTV monitor Introduced at the end of 2009.

Figure 1.5   Sony digital cinema 3D camera(s) and rig. (Courtesy: A. Siliphant. http://www.ana-crhome.com/)

can be immediately plugged in and used with a 3D screen. Sony states that 3D sets will compose 30 to 50 percent of all the TVs it sells in its 2012 fiscal year. Figure 1.5 depicts a modified Sony 3D camera rig by Anachrome 3D that can shoot "near uncompressed" 1920×1080p, when connected to an external memory pack.

In general a technology introduction process spans three phases:

- Stage 1: Time when one can answer affirmatively the question: Is the technology available to support a given service?

- Stage 2: Time when one can answer affirmatively the question: Has a standard emerged to support widespread deployment?
- Stage 3: Time when one can answer affirmatively the question: Is the technology inexpensive enough to foster large-scale adoption by the end-user?

In reference to 3DTV, we find ourselves at some midpoint in Stage 1. However, there are several retarding factors that could hold back short-term deployment of the technology on a broad scale, including deployment and service cost (overall status of the economy), standards, content, and quality.*† Formal industry advocates are beginning to emerge. For example, the 3D@Home Consortium was formed in 2008 with the mission to speed the commercialization of 3D into homes worldwide and provide the best possible viewing experience by facilitating the development of standards, road maps, and education for the entire 3D industry—from content and hardware and software providers to consumers [3DA201001]. It will probably be around 2012 by the time that there will be an interoperable standard available in consumer systems to handle all the delivery mechanisms for 3DTV [MER200902].

In the meantime an array of operators are planning 3DTV services. We already mentioned ESPN, Discovery Communications, and British Channel 4. Others are expected to join the ranks over time. For example, Welho, the cable broadcaster based in Finland, has begun testing 3DTV with customers located in the Helsinki area; the company is looking to offer a full 3DTV channel at some point during 2010.

There is also interest to develop technologies and core applications enabling end-to-end all-3D imaging mobile phones; the aim of these efforts is to have all fundamental functions of the phone—media display, user interface (UI), personal information management (PIM) applications—realized in 3D. Mobile phones with 3D capabilities do

---

* Poor-quality 3D content can strain the eye and make the viewer nauseous.
† Some claim that while it is "easy to strap cameras together and acquire 3D content … and 3D content has been around for a long time, the art and craft of creating quality 3D is only grasped by a small group of 3D cinematographers" [CHI200901]. While assertions like this might be debatable, it is clear that if an entire industry and revenue stream has to emerge, this situation has to change very quickly.

already exist. Since wearing glasses is generally impractical in this context, autostereoscopic displays with two (or only a few) views are used. Besides TV and mobile phone applications for 3D, other moving-picture-based applications are emerging in the fields of situational awareness, battlefield management, medical image viewing, scientific data visualization, molecular modeling, viewing complex mathematical surfaces, visualization of biological and chemical structures, games, sports stadium displays, and education, among others. For an example, a project called Tele-immersive Environment for Everybody (TEEVE) connects off-the-shelf 3D cameras to a PC; a renderer* is used to project the virtual interactions on a big screen monitor, creating a real-time virtual 3D effect. This is similar to Web conferencing but with a 3D/virtual reality capability [GAN200901]. The title of this other reference is self-explanatory [FER200901]. The list of 3D imaging applications (static imaging instead of moving-picture-based content) is fairly extensive [MIN199401].

Transport of 3DTV signals likely will make use of some of the DVB delivery infrastructure, such as appropriate encapsulation of transport streams for transport/transmission over, say, satellite links. With a general movement toward Internet protocol (IP)-based services it makes sense to develop this technology with IP television (IPTV) mechanisms in mind (for example, see [MIN200801]). However, the IP network infrastructure will likely have to be dedicated to this application to obtain the desired quality of service (QoS) and quality of experience (QoE), rather than the open Internet. Given that the 3DTV service is forward looking, it would make sense to consider IPv6 from the get-go; IPv6 header compression will likely be used in this context.†

---

* A render is a software application (package) that supports rendering. 3D/2D rendering is the process of creating a 2D image or animation from a scene, especially for a computer-generated 3D scene. It is a computer graphics process of automatically converting 3D wire-frame models into 2D images with 3D photorealistic effects on a computer. Computer graphic rendering is used, for example, to simulate the appearance of lighted objects for design applications in architecture, for animation and simulation in the entertainment industry, and for display and design in the automobile industry [WES200201].

† IPv6 compresses better than IPv4. It compresses down to 2 bytes. Since IPv6 is a fixed header, it will be even faster when compressing and transmitting since the size is predictable. This is very important for the mobile devices to reduce the overhead on spectrum.

Overall the following 3D applications are being investigated by various researchers and advocates:

- Real-time 3D video (3DV) communication
- 3DV playback of downloaded and stored content
- 3DV display of broadcasted content (i.e., 3DTV, via DVB, DVB-H)
- 3DV display of streamed content (i.e., via mobile phone line, WLAN)
- 3DV recording

In this text we focus almost exclusively on 3DTV, but mention 3DV in the playback and cell phone environment as appropriate. It should be noted in closing this introductory exposition on 3DTV that there currently are two camps in this space: those who wish to pursue the stereoscopic approach and those who advocate the autostereoscopic approach.

- The stereoscopic approach follows the cinematic model, is simpler to implement, can be deployed more quickly (including the use of relatively simpler displays), can produce the best results in the short term, and may be cheaper in the immediate future, but the limitations are the requisite use of accessories (glasses), somewhat limited positions of view, and physiological and/or optical limitations, including possible eyestrain.
- The autostereoscopic approach is more complex to implement, will require longer to be deployed (including the need to develop relatively more complex displays and more complex acquisition/coding algorithms), and so may be more expensive in the immediate future, but it can produce the best results in the long term, such as accessories-free viewing, multiview operation allowing both movement and different perspective at different viewing positions, and better physiological and/or optical response to 3D.

Even within the stereoscopic approach there are variances on how to capture and transmit two images, whether this should be via two separate images handled distinctly but by well-established algorithms, or via some more elegant one-image-plus-metadata means. These observations from the industry (from the 3D4YOU European initiative) provide food for thought:

The conventional stereoscopic concept entails two views: it relies on the basic concept of an end-to-end stereoscopic video chain, that is, on the capturing, transmission and display of two separate video streams, one for the left and one for the right eye. [Advocates for the autostereoscopic approach argue that] this conventional approach is not sufficient for future 3DTV services. The objective of 3DTV is to bring 3D imaging to users at home. Thus, like conventional stereo production and 3D Cinema, 3DTV is based on the idea of providing a viewer with two individual perspective views—one for the left eye and one for the right eye. The difference in approaches, however, lies in the environment in which the 3D content is presented. While it seems to be acceptable for a user to wear special glasses in the darkened theatrical auditorium of a 3D Digital Cinema, most people would refuse to wear such devices at home in the communicative atmosphere of their living rooms. Basically, auto-stereoscopic 3D displays are better suited for these kinds of applications. Such displays apply optical principles such as diffraction, refraction, reflection and occlusion to directly steer the individual stereoscopic views to different locations in space. Most of today's auto-stereoscopic 3D displays are based on an approach where two or more perspective views are simultaneously presented on the screen in a column-interleaved spatial multiplex.

In this context the main challenge of 3DTV is to cover the whole range of today's and next-generation 3D displays including conventional stereo projection and displays with shuttered or polarized glasses as well as different of auto-stereoscopic displays tracked or non-tracked auto-stereoscopic single-user displays serving the user with two views only or auto-stereoscopic multi-user displays providing multiple stereo views for supporting more than one user with the stereo content. Today, the number of views that is supported by commercially available 3D displays suitable for 3DTV applications ranges from $M = 2$ to $M = 10$, but it can be expected that it will be much more in future. This requires an adaptation process by which a given number N of views from the production side (e.g., $N = 2$ in case of a stereo production) is converted to the M views needed by a particular 3D display. Obviously, to enable interoperability of a future 3DTV distribution system, this adaptation process has to be done at the receiving side . . .

To meet these requirements, the 3D4YOU concept for 3DTV is based on a data representation format known as "video-plus-depth." [KAU200801]

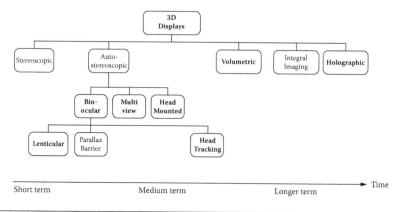

Figure 1.6   Various 3DTV (display) systems.

Figure 1.6 depicts various possible 3DTV display systems. As noted, **stereoscopic 3D** refers to two video streams (or photographs) taken from slightly different angles that appear three-dimensional when viewed together; this technology is likely to see the earliest implementation using specially equipped displays that support polarization. **Autostereoscopic**, also as noted, describes 3D displays that do not require glasses to see the stereoscopic image (using lenticular or parallax barrier technology). Whether stereoscopic or autostereoscopic, a 3D display (screen) needs to generate parallax that in turn creates a stereoscopic sense. A **multi-viewpoint 3D system** is a system that provides a sensation of depth and motion parallax based on the position and motion of the viewer; at the display side, new images are synthesized based on the actual position of the viewer. **Integral imaging (holoscopic imaging)** is a technique that provides autostereoscopic images with full parallax by using an array of microlenses to generate a collection of 2D elemental images; in the reconstruction/display subsystem, the set of elemental images is displayed in front of a far-end microlens array. **Holography** is a technique for generating an image (hologram) that conveys a sense of depth but is not a stereogram in the usual sense of providing fixed binocular parallax information; holograms appear to float in space and they change perspective as one walks left and right; no special viewers or glasses are necessary (note, however, that holograms are monochromatic). **Volumetric/hybrid holographics** are systems that use geometrical principles of holography, in conjunction with other volumetric display methods;

they are primarily targeted, at least at press time, to the industrial, scientific, and medical (ISM) communities.

3DTV based on stereoscopy will likely see earlier deployment compared with other technological alternatives. Holography and integral imaging are relatively newer technologies compared to stereoscopy; holographic and integral imaging 3DTV may be feasible late in the decade. Holography and integral imaging ideally provide true full-parallax 3D displays. Stereoscopy, holography, and integral imaging each has its own distinct features, advantages, and limitations. This text focuses principally on stereoscopy. Figure 1.7 (based in part on the 3D4YOU concept) depicts what a longer-term system (a multi-viewpoint system) might look like. This text will guide the interested players and planners through the available choices.

Figure 1.8 gives some context to the system-level complexity of building a national infrastructure of 3DTV providers, where such providers are properly positioned to deliver a commercial-grade-quality bundle of multiple 3DTV content channels to premium-paying customers. This figure depicts an IPTV-based delivery model, but other transport mechanisms (such as cable TV, satellite DTH, and so forth are possible and/or desirable). While there is a lot of academic interest in subelements of the overall system, the paying public only sees a system-level view with multiple content providers, multiple aggregators, and multiple choices. Service providers see multiple content providers/aggregators and a large multitude of customers. A "system" that can meet these service expectations is needed if 3DTV/3D has to graduate from its legacy of the past 50-plus years.

1.3 Course of Investigation

The issues highlighted in this chapter are covered extensively in this book. This is the first practical, nonacademic book on the topic, and it offers an early practical view of this technology for interested planners, researchers, and engineers.

3DTV service stakeholders need to consider (obviously beyond the basic technological building blocks) a system-level view of what it takes to deploy an infrastructure of 3DTV providers, where such providers are positioned to properly deliver a commercial-grade-quality bundle of multiple 3DTV content channels to paying customers.

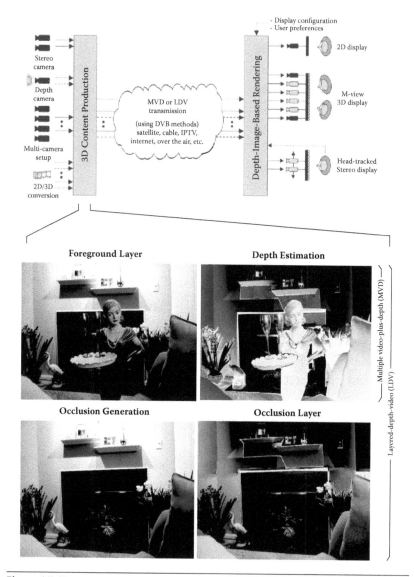

Figure 1.7 (Please see color insert following page 160)   Example of possible long-term 3DTV system.

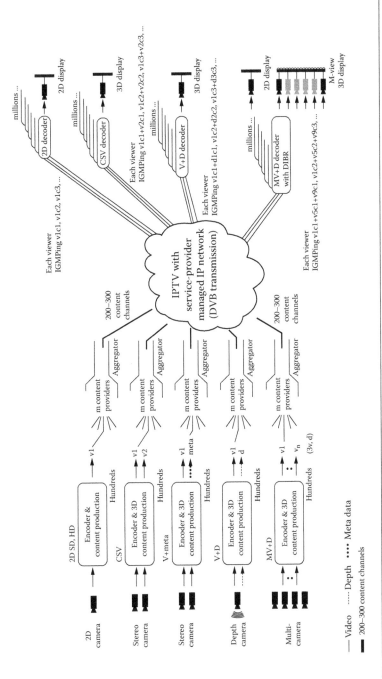

Figure 1.8 Complexity of a commercial-grade 3DTV delivery environment using IPTV.

While there is a lot of academic interest in subelements of the overall system, as depicted in illustrative fashion in Figure 1.9, the paying public and the service providers ultimately must see a system-level view of the delivery apparatus.

This text, therefore, takes such a system-level view. The contribution of this text, we hope, is to have the reader see the picture in the heavy border of Figure 1.10 and not only the elements of Figure 1.9.

Fundamental visual concepts supporting stereographic perception of 3DTV are reviewed in Chapter 2. Stereographic perception is applied to the requirements for supporting 3DTV in Chapter 3. 3DTV technology and digital video principles are discussed in Chapter 4. Elements of an end-to-end 3DTV system are covered in Chapter 4, including a discussion of transmission technologies and advocacy for IPv6 based 3DTV IPTV. End-user devices are addressed in Chapter 5, and a press-time overview of the industry is provided in Chapter 6.

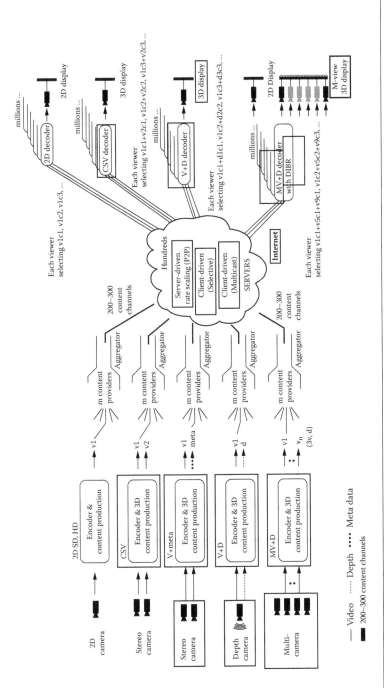

Figure 1.9 Component-level view of 3DTV delivery environment (Internet infrastructure).

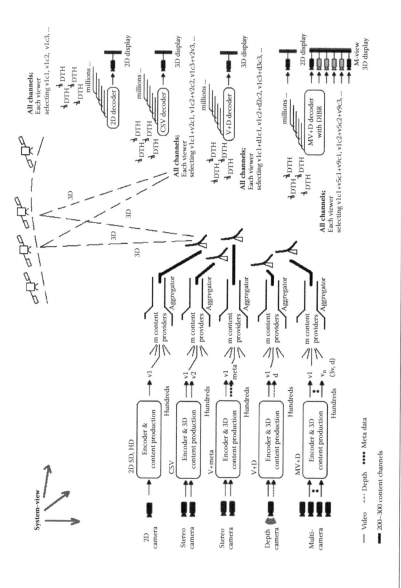

Figure 1.10 System-level view of 3DTV delivery environment (satellite infrastructure).

References

[3DA201001] The 3D@Home Consortium, http://www.3dathome.org/

[3DC200901] 3D Consortium and the Ultra-Realistic Communication Forum, Joint special lectures, All Japan will win the global 3D business Competition, 17 September 2009, Fun Theater in Headquarters Mirai-Kenkyusho of NAMCO BANDAI Games, Higashi-Shinagawa.

[3DM200901] 3D Media Workshop at HHI, Berlin, 15–16 October 2009.

[3DP200802] 3DPHONE, Project no. FP7–213349, Project title: All 3D Imaging Phone, 7th Framework Programme, Specific Programme "Cooperation," FP7-ICT-2007.1.5—Networked Media, D5.1 Requirements and specifications for 3D video, 19 August 2008.

[3DT200701] 3DTV, D26.2 Technical Report #2 on 3D Time-varying scene capture technologies, 7 March 2007, TC1 WP7 Technical Report 2. Ed. C. Theobat, Project Number 511568, Project Acronym 3DTV.

[ALR200901] G. Alregib, Immersive communications: Why now? *IEEE COMSOC MMTC E-Letter* (IEEE Multimedia Communications Technical Committee E-Letter) 4, 3, April 2009.

[ANA200901] A. Silliphant, http://www.anachrome.com/

[AUT200801] Autodesk, Stereoscopic Filmmaking Whitepaper: The Business and Technology of Stereoscopic Filmmaking, 2008, Autodesk, Inc., 111 McInnis Parkway, San Rafael, CA 94903.

[BAR200901] Barco NV, Media and Entertainment Division, "Digital Cinema, Sharing the Digital Experience," Marketing Brochure, 2009, Noordlaan 5, 8520 Kuurne, Belgium, www.barco.com

[BAT200901] J. Bates, 3D HD: The next big entertainer, money maker, *Via Satellite Magazine*, 20ff, September 2009.

[CEA200901] Consumer Electronics Association (CEA), press release, CEA's Industry Forum delivers economic analysis, retail strategy, green trends and industry advice, *Earth Times Online*, 22 October 2009.

[BBC200901] BBC, Sony shows off 3DTV technology, 2 October 2009, http://news.bbc.co.uk

[CHI200901] C. Chinnock, 3D Coming Home in 2010, 3D@Home White Paper, 3D@Home Consortium, www.3Dathome.org

[CHR200601] L. Christodoulou, L.M. Mayron, H. Kalva, O. Marques, and B. Furht, 3D TV Using MPEG-2 and H.264 view coding and autostereo-scopic displays, International Multimedia Conference archive, *Proceedings of the 14th annual ACM International Conference on Multimedia*, Santa Barbara, CA, 2006.

[DOS200801] C. Dosch and D. Wood, Can we create the "holodeck"? The challenge of 3D television, *ITU News Magazine*, no. 9, November 2008.

[DVB201001] Digital Video Broadcasting Project (DVB), Online website material regarding the launch of the DVB 3D TV kick-off workshop, January 2010.

[FEH200401] C. Fehn, Depth-image-based rendering (DIBR), compression, and transmission for a new approach on 3DTV, in *Stereoscopic Displays and Virtual Reality Systems XI*, A.J. Woods, J.O. Merritt, S.A. Benton, and M.T. Bolas, Eds., Bellingham, WA, Proceedings of the SPIE, 5291, 2004.

[FER200901] J. Fermoso, Hong Kong filmmakers to use 3D technology in porn movie, *Wired Magazine Online*, 29 January 2009.

[GAN200901] P. Ganapati, 3D conferencing system allows for virtual light saber duels, *Wired Magazine Online*, 18 June 2009.

[HEW200901] C.T.E.R. Hewage and S. Worrall, Robust 3D video communications, *IEEE COMSOC MMTC E-Letter* 4, 3, April 2009.

[HTH200701] "High Definition Holographic Television," Online Page, 2007, HDHTV.com, Virtual Search, 13033 Ridgedale Drive #140, Minnetonka, MN, 55305-1807.

[ISS199901] I. Sexton and P. Surman, Stereoscopic and autostereoscopic display systems, *IEEE Signal Processing* 16, 3, 85, May 1999.

[ITU200801] ITU-R Newsflash, ITU journey to worldwide "3D television" system begins, Geneva, 3 June 2008.

[JOH200901] C. Johnston, Will new year of 3D drive lens technology? *TV Technology Online*, 15 December 2009.

[KAU200801] P. Kauff, M. Müller, et al., ICT-215075 3D4YOU, deliverable D2.1.2: Requirements on post-production and formats conversion, August 2008.

[LON200801] T. Long, April 8, 1953: Hollywood finally catches 3D fever, *Wired Magazine Online*, 8 April 2008.

[MAE202001] R. La Maestra, 3D TV at CES 2020—Was it Actually Like HD a Decade Ago? (Part 3), HDTV Magazine, February 24, 2020.

[MER200801] R. Merritt, CEA to set 3DTV standard, *EE Times India*, 25 July 2008.

[MER200901] R. Merritt, Incomplete 3DTV products in CES spotlight HDMI upgrade one of latest pieces in stereo 3D puzzle, *EE Times*, 23 December 2009.

[MER200902] R. Merritt, SMPTE to kick off 3DTV effort in June—Task force calls for single home mastering spec, *EE Times*, 13 April 2009.

[MIN199401] D. Minoli, *Imaging in Corporate Environments—Technology and Communication*, McGraw-Hill, New York, 1994.

[MIN200801] D. Minoli, *IP Multicast with Applications to IPTV and Mobile DVB-H*, Wiley/IEEE Press, Hoboken, NJ, 2008.

[ONU200601] L. Onural, A. Smolic, et al., An assessment of 3DTV technologies, *2006 NAB BEC Proceedings*, 456 ff.

[ONU200701] L. Onural, Television in 3-D: What are the prospects? *Proceedings of the IEEE*, 95, 6, 1143, 2007.

[ONU200801] L. Onural and H.M. Ozaktas, Three-dimensional television: From science-fiction to reality, in *Three-Dimensional Television: Capture, Transmission, Display*, H.M. Ozaktas and L. Onural, Eds., Springer Verlag, New York, 2008.

[REA200901] M. Reardon, 3D is coming to a living room near you, CNET Online, 15 January 2009.

[ROS200801]. B. Rosenhahn, Ed., D26.3 Time-varying scene capture technologies, Project Number 511568, Project Acronym 3DTV Initiative Title: Integrated three-dimensional television—Capture, transmission and display, March 2008, TC1 WP7 Technical Report 3.

[SAV200901] V. Savov, LG expects to sell 3.8 million 3D LCDs by 2011, Partners with Korean Broadcaster SkyLife, http://www.engadget.com, 15 December 2009.

[SHI200901] A. Shilov, Blu-ray disc association finalizes stereoscopic 3D specification: Blu-ray 3D spec finalized: New players incoming, *xbitslabs Online*, 18 December 2009, http://www.xbitlabs.com.

[SOE200901] R. So-eui and K. Wills, LG sets 3D TV target to offer new lineup in 2010, Reuters (Seoul), 15 December 2009.

[SMP200901] W. Aylsworth, New SMPTE 3D home content master requirements set stage for new market growth, *National Association of Broadcasters*, April 17, 2009, Las Vegas, NV.

[SOB200901] K. Sobti, 3D comes to the living room with Blu-ray 3D, *Thinkdidit.com Online*, 21 December 2009.

[STA200801] M. Starks, *Digital 3D projection*, 3DTV Corp., Springfield, OR, 2008.

[STA200901] M. Starks, "SPACESPEX™ Anaglyph—The Only Way To Bring 3DTV To The Masses," Online article, 2009.

[SZA201001] G. Szalai and J. Hibberd, ESPN, Discovery launching 3D television networks, 5 January 2010, *The Live Feed*, http://www.thrfeed.com

[TAR200901] G. Tarr, Technology poised to become "major force" in home entertainment, twice, *Video Business*, 23 February 2009.

[WES200201] H.B. Westlund and G.W. Meyer, The role of rendering in the competence project in measurement science for optical reflection and scattering, *J. Res. National Institute of Standards Technology* 107, 3 (May–June 2002).

Bibliography

3DTV only

H.M. Ozaktas and L. Onural, Eds., *Three-Dimensional Television: Capture, Transmission, Display*, Springer Verlag, New York, 2008.

B. Javidi and F. Okano, Eds., *Three-Dimensional Television, Video and Display Technology*, Springer Verlag, New York, 2002.

O. Schreer, P. Kauff, and T. Sikora, Eds., *3D Videocommunication: Algorithms, Concepts and Real-Time Systems in Human-Centered Communication*, John Wiley & Sons, New York, 2005.

D. Minoli, 3DTV Content Capture, Encoding and Transmission, Wiley, New York, 2010.

2

SOME BASIC FUNDAMENTALS
OF VISUAL SCIENCE

3D depth perception can be supported by 3D display systems that allow the viewer to receive a specific and different view for each eye; such a stereopair of views must correspond to the human eye positions, thus enabling the brain to compute the 3D depth perception. In this chapter we explore some basic key concepts of visual science that play a role in 3DTV. The concepts of stereoscopic vision, parallax, convergence, and accommodation are covered, among others. A more detailed discussion for readers interested in some of the theory is included in Appendix 2A. These concepts are then applied to the 3DTV environment in Chapter 3.

2.1 Stereo Vision Concepts

2.1.1 Stereoscopy

Stereo means "having depth, or three dimensions" (the term is used as a prefix to describe or, as a contraction, to refer to stereographic or stereoscopic artifacts or phenomena); it refers to an environment where two inputs combine to create one unified perception of three-dimensional space. **Stereo vision** (also **stereoscopic vision** or **stereopsis**) is the process where two eye views combine in the brain to create the visual perception of one three-dimensional image; it is a by-product of good binocular vision and is the process of vision wherein the separate images from two eyes are successfully combined into one three-dimensional image in the brain. Humans (perhaps even some animals) naturally view their environment in three dimensions; both eyes see a single scene from slightly different positions and the brain then "calculates" the difference and "reports" the third dimension. **Stereographs** (also **stereograms** or **stereopairs**) refer to two images

Figure 2.1   Classical View-Master 3D viewer.

made from different points of view that are side by side. When viewed with a special viewer, the effect is reasonably similar to seeing the objects in reality. Charles Wheatstone invented the first stereoscopic viewer for 3D viewing of stereo pairs in the 1830s. The stereoscopic viewer was based on two pictures and mirrors [3DA201001]. These viewers eventually became popular as the well-known Victorian parlor "stereoscopes." The View-Master toys, which are similar in concept, are still in production today (see Figure 2.1).

Stereoscopy can be defined as any technique that creates the illusion of depth of three dimensionality in an image. **Stereoscopic** (literally "solid seeing") is the term to describe a visual experience having visible depth as well as height and width. The term may refer to any experience or device that is associated with binocular depth perception. **Stereoscopic 3D** (S3D) refers to two photographs, videos, or films taken from slightly different angles that appear three dimensional when viewed together. A **plano-stereoscopic** image is a projected stereoscopic image that is made up from two planar images. **Autostereoscopic** describes 3D displays that do not require glasses to see the stereoscopic image.[*]

Stereogram is a general term for any arrangement of left-eye (LE) and right-eye (RE) views that produces a three-dimensional result that may consist of:

- Simulcasting of two full resolution images
- A side-by-side or over-and-under pair of images

---

[*] Multiview autostereoscopic displays based on parallax barriers or lenticules are sometimes called parallax panoramagram displays.

- Superimposed images projected onto a screen
- A color-coded composite (anaglyph)
- Lenticular images
- A vectograph
- Alternate projected LE and RE images that fuse by means of the persistence of vision

Stereoplexing (stereoscopic multiplexing) is a mechanism to incorporate information for the left and right perspective views into a single information channel without expansion of the bandwidth.

2.1.2 Binocular Depth Perception and Convergence

Binocular is a term to describe an event involving both eyes at once, for example, binocular vision.* **Binocular disparity** is the difference between the view from the left and right eyes. **Binocular depth perception** is the ability (as a result of successful stereo vision) to visually perceive three-dimensional space; namely, the ability to visually judge relative distances between objects, a visual skill that aids accurate movement in three-dimensional space. Many 3D display systems currently envisioned for early 3DTV production are based on the stereoscopic technique to utilize binocular disparity as a depth cue. **Convergence** is the ability of both eyes to turn inwards together, enabling both eyes to be looking at the exact same point in space, as depicted in Figure 2.2. Convergence is essential to maintaining attention and single vision. (For good binocular skills one also needs to be able to look further away; this is called divergence.) Sustained ability to make rapid convergence† and divergence movements are vital to everyday vision. Most 3D display systems use stereoscopic techniques to stimulate the perception of binocular disparity and binocular convergence [BRO200701].

---

* The term binocular stereopsis (two-eyed solid seeing) is used in some psychology books for the depth sense more simply described as stereopsis.
† The term has also been used to describe the movement of left and right image fields or the rotation (toe-in) of camera heads.

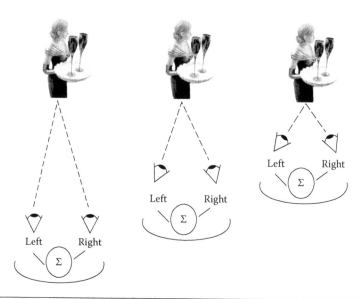

Figure 2.2 (Please see color insert following page 160)   Convergence.

2.1.3 Cyclopean Image

A Cyclopean image is a single mental image of a scene created by the brain by combining two images received from the two eyes; this is also known as fusion—the merging of the two separate views of a stereo pair into a single 3D image (Figure 2.3). The fused image generated from the two stereo images becomes a perception in the human brain that is qualitatively distinct from the two monocular components.

2.1.4 Accommodation

Accommodation (change in ocular focus) and **vergence** (change in ocular alignment) are two ocular motor systems that interact with each other to provide clear binocular vision. Accommodation is the process by which changes in the dioptric power of the crystalline lens cause a focused image to be formed at the fovea. Vergence can be defined as the movement of two eyes in opposite directions. While retinal blur drives accommodation as a reflex, retinal disparity changes accommodative position through the convergence-accommodation (or simply vergence-accommodation, VA) cross-link. Similarly, while retinal disparity primarily drives the vergence system, a change in retinal blur alters vergence through the accommodative-convergence (AC)

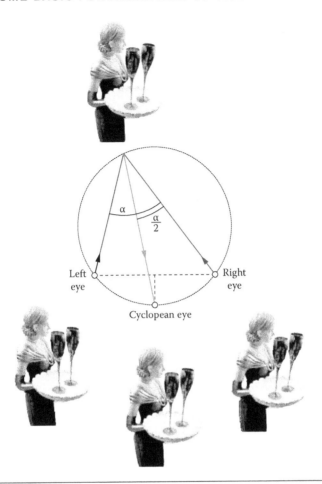

Figure 2.3 (Please see color insert)   Cyclopean image.

cross-link. Under normal binocular viewing conditions, accommodation and vergence interact with each other through reciprocal cross-
link interactions where optically stimulated accommodation evokes
convergence and disparity stimulated vergence evokes accommodation.
The magnitudes of these interactions are quantified as AC/A (ratio
of accommodative convergence to accommodation) and CA/C (ratio
of convergence-induced accommodation to convergence). Although
much information is known on the individual response dynamics of
blur accommodation and disparity vergence, relatively little is known
about the cross-linkages AC and VA. VA represents the unique situation where a stimulus to vergence (retinal disparity) drives a change
in accommodation [SUR200501].

In summary, accommodation is the refocusing of the eyes as their vision shifts from one distance plane to another. The **accommodation/(con)vergence relationship** is the learned relationship established through early experience between the focusing of the eyes and verging of the eyes when looking at a particular object point in the visual world. An **accommodation-(con)vergence link** is the physiological link that causes the eyes to change focus as they change convergence, a link that has to be overcome in stereo viewing since the focus remains unchanged on the plane of the constituent flat images. Binocular disparity, blur, and proximal cues drive convergence and accommodation. Disparity is considered to be the main vergence cue and blur the main accommodation cue [HOR200901].

2.2 Parallax Concepts

2.2.1 Parallax

Parallax is the distance between corresponding points in the left- and right-eye images of a plano-stereoscopic image. **Parallax angle** is the angle under which the optical rays of the two eyes intersect at a particular point in the 3D space. Hence, binocular parallax is the apparent change in the position of an object when viewed from different points (for example, from two eyes or from two different positions); in slightly different words, it is an apparent displacement or difference in the apparent position of an object viewed along two different lines of sight or the differences in a scene when viewed from different points.

In stereoscopic 3D, parallax is often used to describe the small relative displacements between homologues. **Homologues** are identical features in the left and right image points of a stereo pair. The spacing between any two homologous points in a view is referred to as the **separation** of the two images; separation can be employed in determining the correct positioning of the images when displayed as a stereo pair—but note that the separation of the two images varies according to the apparent distance of the points. Nearby objects have a larger parallax than more distant objects when observed from different positions; because of this feature, parallax can be used to determine distances: parallax is measured by the angle or semiangle of inclination between those two lines. Because the eyes of a person are in different positions on the head, they present different views simultaneously. This is the basis of stereopsis, the process by which the brain exploits the parallax due to the different views from the eye to gain depth perception and estimate distances to objects. To further illustrate the concept of parallax, note that parallax error can be seen when taking photos with twin-lens reflex cameras and those including viewfinders (in such cameras, the eye sees the subject through different optics than the one through which the photo is taken). See Figure 2.4. The application of parallax to 3D is depicted in Figure 2.5. More discussion and pictorial diagrams on these concepts are included in Appendix 2A for the reader looking for an analytical formulation of the issue.

Parallax budget is the range of parallax values (from maximum negative to maximum positive) that is within an acceptable range for comfortable viewing. Stereopsis is the binocular depth sense; it is the blending of stereo pairs by the human brain. It can be described as the physiological and mental processes of converting the individual LE and RE images seen by the eyes into the sensation and awareness of depth in a single three-dimensional concept. The parallax view is what gives people a sense of visual depth, or at least the illusion of depth.

The **angular resolution** determines the smallest angle between independently emitted light beams from a single screen point. It can be calculated by dividing the emission range with the number of independently addressable light beams emitted from a screen point.

Binocular parallax and convergence occur when the left and right eyes receive a horizontal displacement by an amount corresponding to the depth distance. Stereoscopic pictures based on this principle have been used for quite some time, even before the invention of photography. As photographs became available, improvements were made on stereoscopes and binocular 3D principles in the second half of the 19th century. The early 1900s saw the invention of parallax barriers, integral photography, lenticular lenses, and other stereoscopic displays [JAV200201]. After the invention of movies, stereoscopic images were often shown at theaters during the 1900s. The main means of stereoscopic display has moved over the years from anaglyth to polarization.*

---

* 3D screens of lenticular lenses have also been used.

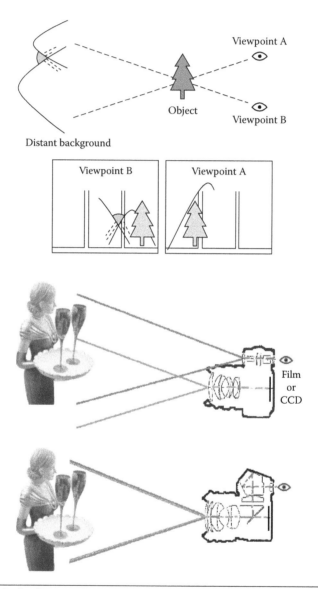

Figure 2.4   Parallax.

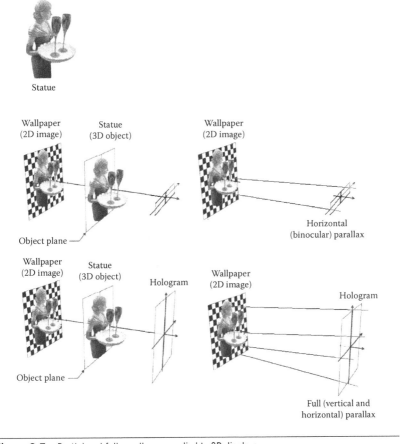

Figure 2.5 Partial and full parallax as applied to 3D displays.

Summarizing this discussion, S3D presents the viewer's eyes with two separate images to create the perception of depth; the displacement between the left and right eye images is the parallax. When the two images are shown simultaneously, one to each eye, parallax produces a retinal disparity causing stereopsis, or the perception of depth. Parallax is a measure of the (horizontal) displacement of an object at the source (e.g. on a display device), while retinal disparity is its effect. In the real world human beings have horizontally separated eyes and therefore we favor horizontal parallax when it comes to the perception of depth (in filmed material, lens distortions, misalignments, or processing can cause vertical parallax; vertical parallax causes eyestrain and should be avoided or corrected). The position in depth (relative to the screen plane) of an object in the scene determines the amount

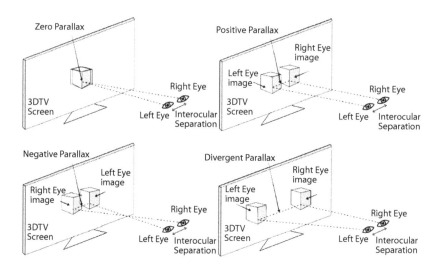

Figure 2.6   Parallax issues as applied to S3D.

and the kind of parallax that it will have in the stereo pair. Making reference to Figure 2.6, an object is said to have zero parallax when it is placed at the same depth as the screen, and causes the two images to lie directly on top of each other while the viewer's eyes converge at the screen plane. Objects with zero parallax appear to be at the same distance as the screen. An object is said to have positive parallax when its parallax is greater than zero parallax—the image of the object presented to the right eye is further to the right than the image presented to the left eye—but less than or equal to the interocular distance. Positive parallax causes the object to appear behind the screen plane and can be infinite in distance. In this latter case—infinite parallax—the parallax is equal to the distance between the eyes (interocular distance), causing the eyes' axes to remain parallel so that the object will appear to be placed at infinity. Past the point of infinite parallax we have divergent parallax, which occurs when the parallax value is greater than the interocular distance, causing the eyes' axes to diverge (the right eye tries to look right, while the left eye looks left). This condition does not occur in real-world vision and requires odd muscle movements. While some viewers claim to be able to adapt to the sensation, it is recommended that strongly divergent parallax be avoided in S3D entertainment productions. Negative parallax occurs when the axes of the viewer's eyes converge in front of the screen, since the image presented to the left eye is further right than the image presented to the right eye, causing the object to appear to be placed between the screen and the viewer. Objects with negative parallax are said to be "in viewer space." Producing stereo images requires two real or virtual cameras. The distance between lenses as referenced by the optical center of each lens is called the interaxial separation (or baseline). Perceived depth is directly proportional to the interaxial separation—that is, as the lenses get farther apart, parallax and the corresponding sense of depth increases. When the interaxial distance is larger than the interocular distance, the effect is called hyperstereoscopy and results in depth exaggeration in the scene. The opposite effect, where the interaxial separation is smaller than the interocular distance, produces a flattening effect on the objects in the scene, and is called hypostereoscopy or cardboarding [AUT200801].

2.2.2 Parallax Barrier and Lenticular Lenses

A **parallax barrier** is a device used on the surface of non-glasses-based 3DTV system with slits that allow the viewer to only see certain vertical columns of pixels at any one time. **Lenticular lenses** are curved optics that allow both eyes to see a different image of the same object at exactly the same time.

2.3 Other Concepts

2.3.1 Polarization

Polarization of light is the arrangement of beams of light into separate planes or vectors by means of polarizing filters; when two vectors are crossed at right angles, vision or light rays are obscured.

2.3.2 Chromostereopsis

Chromostereopsis is the use of color to create the illusion of depth. For significant portions of the population, the images colored in such fashion appear as if they are 3D. Chromostereopsis is a visual perception where two colors of comparable value are placed side by side. When lines (or letters) of different colors are projected or printed, the depths of the lines may appear to be different; lines of one color may "jump out" while lines of another color appear to be recessed [ALL198101].

2.3.3 3D Imaging

Imaging of 3D material can be done via **volumetric representation** or **surface representation** [COS200701]:

> *Volumetric displays* generate imagery from light-emitting, light-scattering, or light-relaying regions occupying a volume rather than a surface in space, as averaged over the display's refresh period. For example, multiplanar volumetric 3D displays produce 3D imagery by projecting a series of 2D planar cross-sections, or slices, of a 3D scene onto a diffuse surface undergoing periodic motion with a period equal to or less than the eye's integration time.

> *Surface representations* (also called multiview) displays reconstruct scenes with a set of one or more pixelized fields, each transmitting in one or more ray directions, or so-called views. The image surface is usually stationary and planar. Multiview displays take many forms: Spatially multiplexed multiview displays include lenticular and parallax-barrier displays, and angle-multiplexed multiview displays include scanned-illumination systems.

2.3.4 Occlusion and Scene Reconstruction

The concept of **occlusion** is used in Chapter 4. Occlusion is the situation of blocking, as when an object blocks light or an object (or person) blocks the view of another object (or person). These issues play a role in free-viewpoint video/TV (FVV/FTV) and multiview systems. Technically, one could say that all regular shadows are a kind of occlusion, but most people reserve the term *occlusion* for a reference to other kinds of light blockings that are not regular shadows from a light. Some 3DTV systems (especially multiview systems) have to address the issue of occlusions.

The 3DTV occlusion issues are similar but not totally identical to those faced in rendering computer-graphics imagery (CGI). Two

common types of occlusions encountered in rendering include the following [BIR200601]:

Ambient Occlusion. Ambient occlusion is a function designed to
darken parts of the scene that are blocked by other geometry
or less likely to have a full view of the sky. One can use ambient occlusion as a replacement or supplement to the shadows
in the fill lights. The main idea behind ambient occlusion is
hemispheric sampling or looking around the scene from the
point of view of each point on a surface. One can imagine how
rays are sampled in all directions from a point being rendered.
The more of these rays that hit an object (instead of shooting off into empty space), the darker the ambient occlusion.
Ambient occlusion can usually have a maximum distance set
for the rays so that only nearby objects will cause the surface
to darken. Ambient occlusion looks around the whole scene
from the point being rendered, and darkens the point based
on how many nearby objects are encountered that might be
blocking light.

Occlusion in Global Illumination. Global illumination (GI) is an
approach to rendering in which indirect light is calculated as
it inter-reflects between surfaces in the scene. GI is different from ambient occlusion, which functions solely to darken
parts of the scene. GI adds light to the scene, to simulate
bounced or indirect light, essentially replacing both fill lights
and their shadows.

In a traditional approach to 3DTV, one has an end-to-end stereoscopic video chain (capturing, transmitting, and displaying two separate video streams, one for each eye), but new approaches are being
advocated, including depth-image-based rendering (DIBR), where
the input is comprised of a monoscopic stream plus a depth stream,
and the output to the display device is a stereoscopic stream. Depth
information indicates the distance of an object in the 3D scene from
the camera viewpoint, typically represented by 8 bits. Depth maps are
applicable to a number of multimedia applications including 3DTV
and free-viewpoint television (FTV). Free-viewpoint TV/video is a
viewing environment where the user can choose his or her own viewpoint. Important issues in support of DIBR-based 3DTV include (1)

3D content generation, (2) coding and transmission, (3) virtual-view synthesis, and (4) 3D display. As noted in this list, new, virtual-view synthesis is a key consideration for these kinds of approaches to 3D. This involves new views from different points of reference; the creation of the new views may or may not entail occlusion. A number of techniques have been developed over the years to address the problem of 3D reconstruction of an object in volumetric or surface representations in the context of *new view synthesis*. Each of these methods has strengths and weaknesses.

2.4 Conclusion

We wrap up this section with a basic glossary of relevant human vision terms (Table 2.1).

Table 2.1   Basic Glossary of Relevant Human Vision Terms

TERM	DEFINITION
Accommodation/convergence conflict	The deviation from the learned and habitual correlation between accommodation and convergence when viewing plano-stereoscopic images
Angular disparity	See parallax angle
Binocular cues	Depth cues that depend on perception with two eyes
Binocular rivalry	Perception conflicts that appear in cases of colorimetric, geometric, photometric, or other asymmetries between the two stereo images
Convergence	The horizontal rotation of eyes or cameras that makes their optical axes intersect at a single point in 3D space; this term is also used to denote the process of adjusting the zero parallax setting (ZPS) in a stereoscopic camera
Corresponding points	The points in the left and right images that are pictures of the same point in 3D space
Crossed disparity	Retinal disparities indicating that corresponding optical rays intersect in front of the horopter or the convergence plane
Cross-talk	Imperfect separation of the left- and right-eye images when viewing plano-stereoscopic 3D content; cross-talk is a physical entity, whereas ghosting is a psychophysical entity
Depth cues	Cues by which the HVS is able to perceive depth
Depth range	The extent of depth that is perceived when a plano-stereoscopic image is reproduced by means of a stereoscopic viewing device
Disparity	The distance between corresponding points on left- and right-eye images
Diplopia	Perception of double images caused by imperfect stereoscopic fusion
Stereoscopic fusion	The process performed by the brain that combines two images—seen respectively by the left and right eyes—into a single, three-dimensional percept
Ghosting	The perception of cross-talk is called ghosting
Horopter	The 3D curve that is defined as the set of points in space whose images form at corresponding points in the two retinas (i.e., the imaged points have zero disparity)
HVS	Human visual system; system by which humans perceive visual cues
Interaxial distance	The distance between the left- and right-eye lenses in a stereoscopic camera
Interocular distance (IO)	The distance between an observer's eyes, about 64 mm for adult.

Continued

Table 2.1 (*Continued*)   Basic Glossary of Relevant Human Vision Terms

TERM	DEFINITION
Image-based rendering	The process of calculating virtual views on the basis of real images and assigned per-pixel depth or disparity maps
Monocular cues	Depth cues that can be appreciated with a single eye alone, such as relative size, linear perspective, and motion parallax
Multiple video-plus-depth (MV+D)	The concept of representing 3D content with multiple video views where a dense per-pixel depth map is assigned to each view
Negative parallax	Stereoscopic presentation where the optical rays intersect in front of the screen in the viewer's space (refers to crossed disparity)
Normalized cross-correlation (NCC)	Correlation measure for detecting the most likely estimate of point correspondences in stereo vision
Panum's fusional area	Small region around the horopter where retinal disparities can be fused by HVS into a single, three-dimensional image
Parallax	The distance between corresponding points in the left- and right-eye images of a plano-stereoscopic image
Parallax angle	The angle under which the optical rays of the two eyes intersect at a particular point in the 3D space
Plano-stereoscopic	Term for describing 3D displays that achieve a binocular depth effect by providing the viewer with images of slightly different perspective at one common planar screen
Point of convergence	3D point where optical axes of eyes or convergent cameras intersect
Plane of convergence	Depth plane where optical rays of sensor centers intersect in case of parallel camera setup
Positive parallax	Stereoscopic presentation where the optical rays intersect behind the screen in the screen space (refers to uncrossed disparity)
Puppet-theater effect	A miniaturization effect in plano-stereoscopic images that makes people look like animated puppets
Retinal disparity	Disparity perceived at the retina of the human eyes
Sensor shift	Horizontal off-center shift of stereo cameras to allow zero parallax settings (ZPS) in a parallel camera setup
Screen space	The region behind the display screen surface; objects will be perceived in this region if they have positive parallax
Stereopsis	The binocular depth sense
Stereoscopic fusion	Ability of human brain to fuse the two different perspective views into a single, three-dimensional image
Stereoscopy	The method of creating a pair of planar stereo images
Stereo window	The cone defined by the camera's focal points and the borders of the screen within the convergences plane

Table 2.1 (*Continued*)   Basic Glossary of Relevant Human Vision Terms

TERM	DEFINITION
Uncrossed disparity	Retinal or camera disparities where the optical rays intersect behind the horopter or the convergence plane
Vieth-Muller circle	See horopter
Viewer space	The region between the viewer and the display screen surface; objects will be perceived in this region if they have negative parallax
Zero parallax setting (ZPS)	Defines the point(s) in 3D space that have zero parallax in a plano-stereoscopic image created, e.g., with a stereoscopic camera. These points will be stereoscopically reproduced on the surface of the display screen

Source: P. Kauff, M. Müller, et al., ICT-215075 3D4YOU, Deliverable D2.1.2: Requirements on post-production and formats conversion, August 2008.

Appendix 2A: Analytical 3D Aspects of the Human Visual System

This appendix provides a detailed description of the human visual system (HVS) and its ability to perceive depth in the 3D world. It is based on a public report of the 3D4YOU project under the ICT (Information and Communication Technologies) Work Programme 2007–2008, a thematic priority for research and development of the European Commission under the specific program "Cooperation" of the Seventh Framework Program (2007–2013) [KAU200801]. We include it here because of the quality, clarity, and pedagogical value it brings to the topic.

Some readers may choose to skip this appendix on first pass; the concepts are defined elsewhere in the text without the formalism used herewith.

2A.1 Theory of Stereo Reproduction

This section identifies key parameters that are used in the remainder of this report, as listed in Table 2A.1 and Table 2A.2. In Section 2A.2.1, some important knowledge about the human visual system (HVS) and its ability to perceive depth in the 3D world is provided. This is followed by Section 2A.2.2, which describes in some detail the geometric basics of stereo perception at 3D displays. Then, Section 2A.2.3 explains the geometry of stereo capturing. Finally, Section 2A.2.4 concludes with the stereo formula and the explanations of stereoscopic 3D distortions that are system immanent in stereo reproduction.

Table 2A.2 Geometry of 3D Stereo Capturing

t_c	Interaxial distance between two stereo cameras
d	Disparity between corresponding points in left- and right-camera images
F	Common focal length of stereo or multiview setup
Z	Absolute depth of a real scene captured by a stereo camera
Z_{conv}	Depth of convergence plane in a parallel stereo camera
h	Symmetric sensor shift of a parallel stereo camera defining the convergences plane
Z_{near}	Nearest scene object in a captured 3D scene
Z_{far}	Farthest scene object in a captured 3D scene
d_{min}	Minimal disparity between left- and right-camera images for a captured 3D scene
d_{max}	Maximal disparity between left- and right-camera images for a captured 3D scene
Δd	Maximal disparity range for a given 3D scene captured by a parallel stereo camera
W_s	Physical width of used image sensor
Δd_{rel}	Maximal disparity range normalized to sensor width (equal to ΔP_{rel})
S_M	Magnification factor between senor and display size

Table 2A.1 Geometry of 3D Stereo Displays

t_e	Interocular distance (eye distance)
Z_v	Perceived depth while watching 3D content at a stereo display
Z_D	Viewing distance while watching 3D content at a stereo display
P	Parallax between left- and right-eye images presented at a plano-stereoscopic display
P_{min}	Minimal parallax between left- and right-eye images for a given 3D content
P_{max}	Maximal parallax between left- and right-eye images for a given 3D content
$Z_{v,near}$	Minimum of perceived depth (nearest scene object) caused by P_{max}
$Z_{v,far}$	Maximum of perceived depth (farthest scene object) caused by P_{min}
α	Parallax angle under which optical rays intersect at observed 3D point
$\Delta\alpha_{max}$	Maximal range of α between near and far scene objects
$\Delta\alpha_{near}$	Maximal range of α between near scene objects and screen surface
ΔP_{max}	Maximal parallax range between near and far scene objects
W_D	Physical width of 3D display
ΔP_{rel}	Maximal parallax range normalized to display width

2A.2 Analytics

2A.2.1 Depth Perception The HVS is able to perceive depth to the accomplishment of the human brain to interpret several types of *depth cues* that can be separated into two broad categories: *monocular* cues and *binocular* cues. Sources of information that require only one eye (such as relative size, linear perspective, or motion parallax) are called *monocular* cues, whereas sources of information that depend on both eyes are called *binocular* cues. Everyday scenes usually contain more than one type of depth cue and the importance of each cue is based on learning and experience. In addition to this, the influence of the different cues on the human depth perception also depends on the relative distances between the observer and the objects in the scene [CUT199701]. Binocular cues are mainly dominant for viewing distances below 10 m (about 30 feet) and hence they are particularly important for 3D television. They are based on the fact that the human eyes are horizontally separated. Each eye provides the brain with a unique perspective view of the observed scene. This horizontal separation—on average approximately 65 mm for an adult—is known as *interocular distance*, t_e. It leads to spatial distances between the relative projections of observed 3D points in the scene onto the left and the right retinas, also known as *retinal disparities*. These disparities provide the HVS with information about the relative distance of objects and about the spatial structure of our 3D environment. It is the retinal disparity that allows the human brain to fuse the two different perspective views from the left and the right eyes into a single, three-dimensional image.

Figure 2A.1 illustrates how this process of *stereoscopic fusion* works in detail. Basically, when looking at the 3D world, the eyes rotate until their optical axes converge (intersect in a single point) on the "object of interest" [LIP198201]. It follows that the *point of convergence* is projected to same corresponding image points on the respective retinas, that is, it does not produce any retinal disparity. The same holds true for all points on the *horopter*—a curve also known as the *Vieth-Muller circle*—which is defined by the fixation point and the nodal points of both eyes. All other points, however, produce retinal disparities whose magnitudes become larger the further away the 3D points are from the horopter.

Disparities that are caused by points in front of the horopter are said to be *crossed*, and disparities that result from points behind the

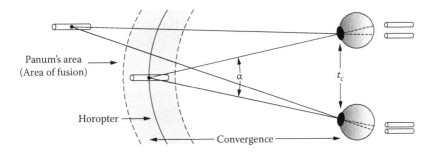

Figure 2A.1 Principle of stereoscopic fusion and retinal disparity.

horopter are therefore called *uncrossed*. As long as the crossed or uncrossed disparities do not exceed a certain magnitude, the two separate viewpoints can be merged by the human brain into a single, three-dimensional percept. The small region around the horopter within which disparities are fused is known as *Panum's fusional area*. Points outside this area are not fused and double images will be seen, a phenomenon that is called *diplopia*.

That these double images usually do not disturb the visual perception is the result of habitual behavior that is tightly coupled with the described convergence process. In concert with the rotation of the optical axes, the eyes also focus (accommodate by changing the shape of the eye's lenses) on the object of interest. This is important for two different reasons. First of all, focusing on the point of convergence allows the observer to see the object of interest clearly and sharply. Second, the perception of disturbing double images, which in principle result from all scene parts outside Panum's fusional area, is efficiently suppressed due to an increasing optical blur [ISM200201].

2A.2.2 Geometry of Stereoscopic 3D Displays   Although particular realizations differ widely in the specifically used techniques, most stereoscopic displays and projections are based on the same basic principle of providing the viewer with two different perspective images for the left and the right eyes. Usually, these slightly different views are presented at the same planar screen. These displays are therefore called *planostereoscopic* devices. In this case, the perception of binocular depth cues results from the spatial distances between corresponding points in both planar views, that is, from the so-called *parallax P*, which in turn induces the retinal disparities in the viewer's eyes.

Thus, the perceived 3D impression depends, among other parameters like the viewing distance, on both the amount and the type of parallax. As shown in Figure 2A.2, three different cases have to be taken into account here:

1. **Positive parallax:** Corresponding image points are said to have *positive* or *uncrossed parallax P* when the point in the right-eye view lies more right than the corresponding point in the left-eye view; see Figure 2A.2(a). Thus, the related viewing rays converge in a 3D point behind the screen, so that the

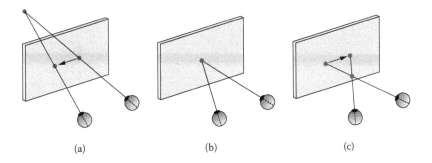

(a)   (b)   (c)

Figure 2A.2 Different types of parallax: (a) positive parallax; (b) zero parallax; (c) negative parallax.

reproduced 3D scene is perceived in the so-called *screen space*. Furthermore, if the *parallax P* exactly equals the viewer's interocular distance t_e, the 3D point is reproduced at infinity. This also means that the allowed maximum of the positive parallax is limited to t_e.

2. **Zero parallax:** With *zero parallax*, corresponding image points lie at the same position in the left- and the right-eye views; see Figure 2A.2(b). The resulting 3D point is therefore observed directly at the screen, a situation that is often referred to as the *zero parallax setting* (ZPS).

3. **Negative parallax:** Conjugate image points with *negative* or *crossed parallax P* are located such that the point in the right-eye view lies more left than the corresponding point in the left-eye view; see Figure 2A.2(c). The viewing rays therefore converge in a 3D point in front of the screen in the so-called *viewer space*.

To a user who is seated at a viewing distance Z_D in front of the screen, an object with horizontal screen parallax P is perceived at depth Z_v (see also Figure 2A.3):

$$Z_v = \frac{Z_D \cdot t_e}{t_e - P} \tag{2.1}$$

Note that $Z_v = Z_D$ holds for $P = 0$. This refers to the ZPS condition; that is, all objects with zero parallax appear at the screen surface. Furthermore, objects with a positive parallax value P appear in the screen space behind the display ($Z_v > Z_D$) as shown in Figure 2A.2(a)

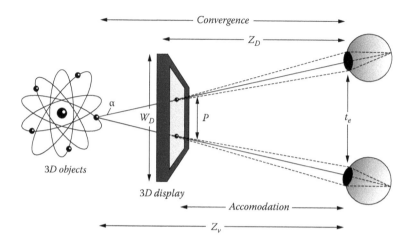

Figure 2A.3   The geometry of viewing 3D displays and the related accommodation-convergence conflict.

and vice versa, objects with a negative parallax value P appear in front of the display in the viewer space $(Z_v < Z_D)$ as shown in Figure 2A.2(c). Thus, from Equation (2.1) one obtains the following relation for the maximal disparity range ΔP that appears at the screen of a stereo-scopic 3D display where $Z_{v,near}$ and $Z_{v,far}$ denote the perceived depth of those scene objects that are most near and, respectively, most far to the viewer.

$$\Delta P = P_{max} - P_{min} = t_e \cdot Z_D \cdot (1/Z_{v,near} - 1/Z_{v,far}) \qquad (2.2)$$

In addition, using the common definition $\alpha = arctan\,(t_e/Z_v)$ for the *parallax angle* under which the optical rays of the two eyes intersect at a particular point in the 3D space, one can define the following to ranges of *angular disparities*:

$$\Delta\alpha_{max} = arctan(t_e / Z_{v,near}) - arctan(t_e / Z_{v,far}) \qquad (2.3a)$$

$$\approx t_e \cdot (1/Z_{v,near} - 1/Z_{v,far})$$

$$\Delta\alpha_{near} = arctan(t_e / Z_{v,near}) - arctan(t_e / Z_D) \qquad (2.3b)$$

$$\approx t_e \cdot (1/Z_{v,near} - 1/Z_D)$$

Analogue to Equation (2.2), $\Delta\alpha_{max}$ denotes the maximal parallax angle between near and far objects that enables a distortion-free fusing of the stereo images, whereas $\Delta\alpha_{near}$ describes the maximal angular disparity between a near object and the screen surface that allows comfortable viewing conditions. Using Equation (2.2) and Equation (2.3a) and assuming that $\Delta\alpha_{max}$ is given, one obtains the parallax range that is maximally allowed at a stereoscopic 3D display:

$$\Delta P_{max} = Z_D \cdot \Delta\alpha_{max} \qquad (2.4)$$

Finally, normalizing Equation (2.4) to the display width W_D results in a relative parallax range ΔP_{rel} that is independent of the particular screen size:

$$\Delta P_{rel} = \Delta P_{max} / W_D = Z_D / W_D \cdot \Delta\alpha_{max} \qquad (2.5)$$

This relation shows that, apart from the physical viewing conditions represented by the ratio Z_D/W_D, ΔP_{rel} mainly depends on a psycho-optical component where $\Delta\alpha_{max}$ describes the maximal parallax under which the stereo images can be fused without visible distortions. As we know from Section 2A.2.1, this parallax angle is unlimited when looking at a real-world 3D scene. In this case the eyes simultaneously *converge* and *accommodate* on the object of interest. As explained, these jointly performed activities allow the viewer to stereoscopically fuse the object of interest and, at the same time, to suppress diplopia (double-image) effects for scene parts that are outside the Panum's fusional area around the focused object. However, the situation is different in stereo reproduction. When looking at a stereoscopic 3D display, the eyes always accommodate on the screen surface, but they converge accordingly to parallax (see Figure 2A.3). This deviation from the learned and habitual correlation between *accommodation* and *convergence* is known as *accommodation-convergence* conflict. It represents one of the major reasons for eyestrain, confusion, and loss of stereopsis in 3D stereo reproduction [IRF200101], [LIP199701], [PAS200201]. It is therefore important to make sure that the maximal parallax angle $\Delta\alpha_{max}$ is kept within acceptable limits—or, in other words, to guarantee that the 3D world is reproduced in rather close distance to the screen surface of the 3D display.

The human factors literature provides a number of different rules of thumb for choosing an appropriate parallax range (see, for example, [CAL193701], [DAC199801], [LIP199701], [LUS194401],

[SMD198801], [YYS199001]). One established rule of thumb is to set $\Delta\alpha_{max}$ to a value of 0.02 (in radiant corresponding to 1.17 degrees or 70 arc minutes). From today's research point of view, this value is too restrictive. It is based on first insights in stereo perception from 1931 [LUS193101], hence it is doubtful if it still holds for state-of-the-art stereo technology. Furthermore it is an open topic in the human factors community whether such a fixed value can globally be applied to the whole screen—especially in the case of large-screen presentations where the observer usually concentrates on a small part of the whole image. All these interrelations are not yet fully understood by the human factors community, and it also seems to differ considerably from one individual observer to the next [JLH200101], [NDB200001], [WDK199301]. Nevertheless the assumption $\Delta\alpha_{max} = 0.02$ represents a conservative estimation and, with it, a simple and safe production rule. It is therefore frequently used in stereo production. As can be seen from Equation (2.5), the relative parallax range ΔP_{rel} furthermore depends on viewing conditions represented by the ratio Z_D/W_D. Thus, roughly, it can be said that content viewed at greater distances or smaller displays allow for larger parallax values [GMM199101], [INO199001]. For stereo projection with medium screen size and viewing distance, it is usual to take a standard ratio of $Z_D/W_D = 1.67$. Hence one can also deduce a related rule of thumb for the relative parallax range ΔP_{rel}:

$$\Delta P_{rel} = Z_D / W_D \cdot \Delta\alpha_{max} = 0.02 / 0.6 = 1 / 30 \qquad (2.6)$$

Another important production general rule refers to the right selection of the angular disparity range $\Delta\alpha_{near}$ from Equation (2.3b), which describes the difference between the parallax angles under which the eyes perceive near objects in front of a screen and far objects at the screen. To achieve pleasant viewing comfort, it is often limited to a value of 20 arc minutes. Note that it is significantly smaller than the $\Delta\alpha_{max}$ (approximately 70 arc minutes) that covers the whole depth range of the scene. $\Delta\alpha_{near}$ is particularly important to control so that near objects do not appear too far from the screen in front of the display, and with it, to decide how the 3D scene is distributed over the viewer and screen space. It is therefore an important production rule for defining the sensor shift h described in more detail in the next section.

2A.2.3 Geometry of Stereo Capturing   The related generation of planar stereoscopic views requires capturing with a synchronized stereo
camera. Because such two-camera systems are intended to mediate
the natural binocular depth cue, it is not surprising that their design
shows a striking similarity with the HVS. For example, the *interaxial
distance* t_c between the focal points of the left- and right-eye camera lenses is usually chosen in relation to the interocular distance t_e.
Furthermore, similar to the convergence capability of the HVS, it
must be able to adapt a stereo camera to a desired convergence condition or zero parallax setting (ZPS), that is, to choose the 3D scene
part that is going to be reproduced exactly on the display screen. As
shown in Figure 2A.4 this can be achieved by two different camera
configurations [WDK199301], [YOO200601]:

1. **"Toed-in" setup:** With the toed-in approach, depicted in
   Figure 2A.4(a), a *point of convergence* is chosen by a joint
   inward-rotation of the left- and the right-eye cameras.
2. **"Parallel" setup:** With the parallel method, shown in
   Figure 2A.4(b), a *plane of convergence* is established by a small
   shift h of the sensor targets.

At first view, the toed-in approach intuitively seems to be the more
suitable solution because it directly fits to the convergence behavior
of the HVS as shown in Figure 2A.2. However, it has been shown in
the past that the parallel approach is nonetheless preferable, because

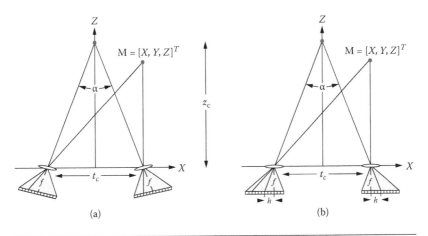

Figure 2A.4   Basic stereoscopic camera configurations: (a) "toed-in" approach; (b) "parallel"
setup.

it provides a higher stereoscopic image quality [WDK199301], [YOO200601]. From today's point of view, the reason for that is clear, and the background seems to be fully understood so far. For the right camera design it is not important that the eyes converge while focusing at a particular part of the scene. In fact, it is more relevant that the two images of a plano-stereoscopic device are displayed at the same planar screen and hence that the two related image planes strictly follow a parallel setup. Hence, with the toed-in configuration, there is a keystone distortion between the image planes of the capturing cameras and the corresponding images planes at the 3D display surface [MIS200401]. This in turn induces both horizontal and unwanted vertical disparities that are known to be one further potential source of eyestrain [IRV200001], [PAS199101]. In contrast, the parallel camera setup only creates the desired horizontal parallax required to perceive the binocular depth effect of plano-stereoscopic devices.

Due to these advantages, the following considerations will mainly concentrate onto the parallel setup from Figure 2A.4(b). Note that the convergent setup from Figure 2A.4(a) can always be converted into a corresponding parallel setup by means of rectification (see also Section 2A.2.4). Thus the following considerations also hold for convergent cameras if rectification is applied beforehand.

In principle, there is a conflict between a parallel camera setup and camera convergence because the optical axes of parallel stereo cameras intersect at infinity. In the setup from Figure 2A.4(b) this conflict is solved be the so-called concept of *sensor shifts*. For this purpose, the sensors of both cameras are shifted horizontally by the same distance h in inverse direction such that the optical rays through the centers of both image planes intersect at the desired point of convergences. It can easily be shown that, using this sensor-shifted parallel setup from Figure 2A.4(b), the following relation holds for disparities d between two corresponding points at the image plane that refer to the same 3D point in the real world:

$$d = t_c \cdot F \cdot (1 / Z_{conv} - 1 / Z) = h - t_c \cdot F / Z \qquad (2.7)$$

Here F and Z_{conv} denote the common focal length of the stereo cameras and the distance of the convergence plane from the camera basis, respectively. Note the analogy to Equation (2.1). Scene parts at $Z = Z_{conv}$ are captured with zero disparity and refer to the ZPS condition. Hence, supposing that no further processing occurs between capture and display, scene objects that have been captured at $Z = Z_{conv}$ will appear at the screen surface of the 3D display. Furthermore, objects in front of the convergence plane $Z = Z_{conv}$ will be captured with negative disparity and will be displayed with negative parallax in the viewer space of the 3D display, and vice versa, objects behind the convergence plane $Z = Z_{conv}$ will cause positive disparities and will be shown in the screen space.

This analogy of capture and display geometry is also shown in Figure 2A.5. It is important to notice that the sensor shift plays an important role in this context. With $h = 0$, the convergence plane $Z = Z_{conv}$ would move to infinity and hence the whole scene would appear in front of the screen surface of the 3D display. Clearly, this is not desired in stereo reproduction. In fact, usually just the opposite is aimed for during proper stereo presentations. Often major parts of the scene are used to appear behind the screen to avoid perception conflicts with the display window. In any case, the scene should be displayed around the screen surface to avoid stereoscopic 3D distortions (see Section 2A.2.4). Thus the sensor shift h is an important parameter to control how the 3D scene is distributed over the viewer and screen space.

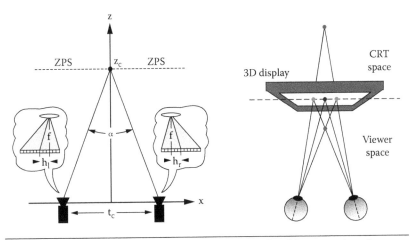

Figure 2A.5 Analogy between capture and display geometry: Parallel camera setup using sensor shifts (left) and perception of 3D scene at stereoscopic display (right).

Similar to Equation (2.2), a maximal disparity range Δd can be defined for stereo capturing by using Equation (2.7):

$$\Delta d = d_{max} - d_{min} = t_c \cdot F \cdot (1/Z_{near} - 1/Z_{far}) \tag{2.8}$$

Furthermore, one can normalize the disparity range Δd from Equation (2.8) by dividing it by the sensor width W_S such that the resulting relative disparity range Δd_{rel} is equal to ΔP_{rel} from Equation (2.5):

$$\Delta d_{rel} = \Delta d / W_S = \Delta P_{rel} \tag{2.9}$$

Thus one can derive the following rule for choosing the right interaxial distance t_c from a given ΔP_{rel}:

$$t_c = \frac{\Delta d}{F \cdot (1/Z_{near} - 1/Z_{far})} = \frac{W_S \cdot \Delta P_{rel}}{F \cdot (1/Z_{near} - 1/Z_{far})} \tag{2.10}$$

Using Equation (2.3a), one can also show that interaxial camera distance t_c is not necessarily equal to the interocular eye distance t_e:

$$t_c = \frac{Z_D \cdot (1/Z_{v,near} - 1/Z_{v,far})}{S_M \cdot F \cdot (1/Z_{near} - 1/Z_{far})} \tag{2.11}$$

Following Equation (2.11), the ratio between t_c and t_e depends on different aspects of capturing and viewing conditions, such as the relation of focal length F to viewing distance Z_D, the so-called magnification factor $S_M = W_D/W_S$, and the ratio of the perceived depth range that can maximally be reproduced by the 3D display to the existing depth range captured in the real 3D world. Section 2A.3 discusses how these formulas are used in stereo production.

2A.2.4 Stereoscopic 3D Distortions   Using Equation (2.1) and Equation (2.7) and taking into account that $P = S_M \times d$, one obtains the following relation between perceived depth Z_v and real depth Z:

$$Z_v = \frac{Z_D \cdot t_e}{t_e - P} = \frac{Z_D \cdot t_e \cdot Z}{S_M \cdot F \cdot t_c - Z \cdot (S_M \cdot h - t_e)} \tag{2.12}$$

From this relation it can be seen that stereoscopic depth reproduction is usually not linear. Linearity only holds if $(S_M \times h - t_e)$ is zero. As

$S_M \times h$ is the screen parallax of points at infinity, one can conclude that depth reproduction is only linear when the camera settings and the projection display setup are such that infinity parallax is equal to the interocular eye distance t_e. Otherwise, stereoscopic distortions occur such that foregrounded scene elements are more elongated than far scene parts ($S_M \times h > t_e$), or vice versa ($S_M \times h < t_e$). This means that only if points at infinity are placed correctly, resulting in parallel viewing rays, will the whole scene fall into correct place, with linear depth reproduction throughout. In this context, it is important to notice that sensor shift h and the convergence plane $Z_{conv} = t_c \times F/h$ are the decisive parameters to control this situation. In principle, stereoscopic 3D distortions can be avoided if $h = t_c/S_M = W_S \times t_c/W_D$. Although it can be supposed that the sensor width W_S is known at the capture side, it still requires some assumptions on the display width W_D. Thus the ideal case can only be achieved for the target display, whereas distortions will occur while viewing the 3D content on another display of different size. The effect of this side condition can be explained more transparently if one rewrites it in the following way:

$$Z_{conv} = \frac{t_c \cdot F}{h} = \frac{t_c \cdot M \cdot F}{t_e} = Z_D \cdot \frac{(1/Z_{v,near} - 1/Z_{v,far})}{(1/Z_{near} - 1/Z_{far})} \quad (2.13)$$

This means that, if h has been chosen correctly, convergence plane Z_{conv} is placed relatively at the same position within the captured scene as the screen of the 3D displays appears in the perceived 3D scene.

When using a convergent camera setup [that is, the toed-in approach from Figure 2A.4(a)] in connection with a subsequent rectification, the sensor shift occurs implicitly. Here, the convergence plane Z_{conv} is already defined by selecting a convergence point in the real 3D scene and by adjusting the convergence angle of the cameras accordingly. As shown in Figure 2A.6, the subsequent rectification process goes along with an implicit sensor shift that keeps the convergence plane at the same position. The situation is somewhat different in the case of using the parallel setup from Figure 2A.4(b) because the convergence plane is now not defined by the setup itself but has to be adjusted independently. In principle, this can be achieved by shifting optical axes and image planes relative to each other. Performing this during capturing, however, requires complicated mechanics moving

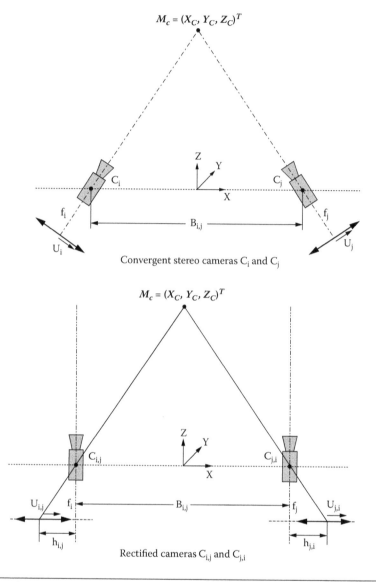

Figure 2A.6   Concept of rectification: convergent camera setup used during capturing and "virtual" parallel setup after rectification.

either the sensor or the lens. As an option, the shifts can be done manually during postproduction. In this case, the scene is captured with a parallel setup using regular cameras without sensor shift (that is, $h = 0$) and the desired sensor shift h is then applied manually to the recorded images during postproduction such that the convergence plane is placed correctly, to Equation (2.13). To this end it has to be concluded that the sensor shift h of a final parallel setup is always unequal to zero, regardless of the particular camera setup used during capturing. In the ideal case, the sensor shifts can be chosen in such a way (either by rectification in a convergent setup or by adding sensor shifts mechanically during capturing or manually during postproduction) that $S_M \times h - t_e$ is at least approximately equal to zero, and such that one obtains an almost linear relation between real and perceived depth in Equation (2.12). Its use for postproduction will be explained in the next section.

2A.3 Workflow of Conventional Stereo Production

Content is critical to the successful commercial launch and penetration of 3D service. Some concepts related to this topic are discussed in this section, which reviews some basic rules of production grammar and production workflows in conventional stereo 3D production. The objective is to learn from conventional stereo production and to reflect these insights for defining the requirements of future 3DTV postproduction.

2A.3.1 Basic Rules and Production Grammar   Conventional stereo production has to respect a number of rules in order to guarantee an appropriate stereoscopic viewing comfort. Some of these rules are independent of the particular scene content, the targeted 3D display platform, and the expected viewing conditions. They mainly refer to technical issues of the capturing device:

Temporal Synchronization: Very precise temporal synchronization between the capturing left- and right-eye cameras is necessary to ensure acceptable 3D quality. With objects in motion, any error in synchronization introduces horizontal and/or vertical alignment errors that might have a negative effect on the stereoscopic viewing comfort.

Colorimetric Symmetry: Asymmetries in luminance, contrast, and color have to be avoided as best as possible. Significant colorimetric differences between the two views of a stereoscopic image pair can lead to *binocular rivalry*. Although the impact of colorimetric differences is not fully investigated and understood yet, there is a common consensus in literature that the two stereo images should be identical, especially in contrast, but with some lower criticality also in luminance and color such that if displayed side by side no discrepancy is visible [PAS199101], [DAC199801]. This requires a suitable colorimetric adjustment of the two cameras during capturing and, if possible, a careful color grading during postproduction.

Geometric Symmetry: Geometric asymmetries such as errors in image height, size, and/or scale, or nonlinear distortions between the left- and right-eye views can result from imperfect stereo setups with nonparallel camera axes, differences in focal length, skewed camera sensors, or optical lenses that have notable radial distortions. Such discrepancies between the two views can lead to visual strain when they exceed a certain threshold. This particularly holds for the toed-in approach from Figure 2A.4(a), which causes a significant keystone distortion. Therefore a subsequent rectification is needed directly during capturing or afterwards during postproduction to compensate for these keystone distortions [WDK199301]. But in some respects it also applies to the parallel setup from Figure 2A.4(b) because parallelism of camera axes as well as lenses and sensor mounting will never be perfect in a real stereo setup [MIS200401]. Geometric distortions must be corrected manually during postproduction or automatically, and if needed, in real time by using calibration data.

Other rules take into account the expected 3D display platform and viewing conditions, often in concert with knowledge about the scene content. One main objective of this production workflow is to choose the right values for the interaxial camera distance t_c and the sensor shift h in dependence on the allowed parallax range ΔP_{rel} and the desired ZPS conditions, respectively:

Interaxial Camera Distance: The interaxial camera distance is usually set in accordance with Equation (2.10) in Section 2A.2.3. For this purpose, the required relative parallax range ΔP_{rel} is selected with dependence on the targeted 3D display or projection platform. As already explained in Section 2A.2, the absolute maximum of ΔP_{rel} is defined by perceptual constraints such as the maximally allowed parallax angle $\Delta \alpha_{max}$ on one hand and the particular viewing conditions on other hand [see Equation (2.5) in Section 2A.2.2]. However, there might be even stronger restrictions from display technology. Almost all current stereoscopic reproduction techniques have a certain amount of cross-talk. It is perceived as ghosting, that is, as double contours, and even a relatively small amount of cross-talk can lead to headaches [MIS200401]. The grade of resulting degradations increases with increasing parallax values. Thus the upper ΔP_{rel} limit of a particular display technology might be much lower than the perceptual limit from Equation (2.5). This especially holds for autostereoscopic displays. In any case, the interaxial camera distance t_c calculated by Equation (2.10) depends on the given scene structure and is usually lower than the interocular eye separation t_e (under normal indoor or outdoor scenes in a range from 30 mm to 70 mm, but might be significantly larger than 70 mm in case of long shots).

Sensor Shifts and Convergence Plane: As explained in Section 2A.2.3, the sensor shift is an appropriate instrument to minimize stereoscopic 3D distortions on one hand and to control how the 3D scene is distributed over viewer and screen space at the display side on other hand. This especially applies to parallel camera setups from Figure 2A.4(b), where the sensor shift is usually chosen independently of the interaxial camera distance t_c. To minimize stereoscopic 3D distortion, h should be set to $h = t_e/S_M$ as discussed in Section 2A.2.4. Similar to the above considerations on the right selection of t_c, this procedure requires some assumption on the 3D display, mainly the targeted display width W_D and its relation to a known sensor width W_S. However, in addition, there might be further content-based constraints to be respected. One example is the avoidance of perception conflicts within the stereo window. Whenever a part of a 3D scene is stereoscopically reproduced in viewer space in front of the screen, it must be ensured that it is not cut off at the borders of the image [KLE199801]. If this happens, the result is a very annoying conflict between two different depth cues: while the binocular depth cue tells the viewer that the object is in front the screen, the monocular cue of interposition at the same time indicates that the object must be behind the screen because it seems to be obscured by the screen surround. Thus, it should in general be ensured that viewer space objects are completely inside of the stereo window. Again, this situation can be controlled by changing the sensor shift during postproduction such that the whole scene content is moved behind the display into the screen space if a conflict of the scene content with the stereo windows occurs.

2A.3.2 Example   Based on this production grammar, some specific workflows have been established by practitioners in 3D stereo production. Depending on the technical equipment used during capturing and postproduction, they might differ in detail, but in principle they follow the same basic rules. The following considerations explain as an example one of the production workflows used by the 3D4YOU partner KUK. Due to known advantages already discussed in Section 2A.2, the stereo production workflow of KUK is strictly based on the parallel camera setup from Figure 2A.4(b). For this purpose, two different types of stereo rigs are used:

90-degree setup: In order to achieve small interaxial distances t_c, the left- and right-eye cameras are mounted on a 90-degree angle, with one camera shooting into a semitransparent mirror and one shooting through it. This technical concept is based on a patent from F. A. Ramsdell that dates back as early as 1947 [RAM194701]. It has been used in many stereo rigs such as Cobalt Entertainment's 3ality stereoscopic 3D camera, Disney's 3D camera, the Hines StereoCam, and the BiClops and Iwerks 870/1570 camera rig [COB200701], [HIN198501].

Side-by-side setup: With a "side-by-side" setup, two cameras are placed next to each other. These systems usually lack some flexibility in adjusting the interaxial distance t_c, as its minimum value is restricted by the width of the thickest part of the camera (which can be either the camera body or the lens). Stereoscopic 3D cameras that use a side-by-side setup are often fully integrated systems that are based on stripped-down digital HD cameras or 35/70 mm film such as the 3DVX3 stereoscopic 3D camera, the Cobalt TS3 camera, the IMAX 3D camera, and the RCS Reality Camera System [21C200701], [COB200701], [IMA200701], [LIG200701], [PAC200701].

The adjustment of interaxial distance t_c to the scene content is usually done by using a special stereoscopic viewfinder. In this viewfinder, the two stereo images are overlaid via two LCD displays, again by using semitransparent mirror technology. One of the displays can be moved until objects or feature points in the near clipping plane are placed in the zero-parallax plane such that the condition $Z_{conv} = Z_{near}$ holds. Then, the interaxial distances t_c of the stereo rig is changed until the relative parallax of objects or feature points in the far clipping plane Z_{far} is equal to the required ΔP_{rel} from Equation (2.6). As already mentioned, ΔP_{rel} is selected case by case and depends on the targeted 3D display platform. Note that this adaptation process is a visualization of the constraint defined in Equation (2.10). Another workflow option for the right adjustment of the interaxial distance t_c is the usage of a complete table framework listing the right t_c value in accordance with Equation (2.10) for all relevant combinations of Z_{near} and Z_{far} and depending on the targeted relative parallax range ΔP_{rel}, the used focal length F, and the sensor width W_S of the used cameras.

References

[21C200701] 21st Century 3D (Online), 2007, http://www.21century3d.com

[3DA201001] The 3D@Home Consortium, http://www.3dathome.org/

[ALL198101] R.C. Allen and M.L. Rubin, Chromostereopsis, *Surv. Ophthalmol.*, 26, 1, 27–27, July–August 1981.

[AUT200801] Autodesk, Stereoscopic Filmmaking Whitepaper: The Business and Technology of Stereoscopic Filmmaking, 2008, Autodesk, Inc., 111 McInnis Parkway, San Rafael, CA 94903.

[BIR200601] J. Birn, Digital Lighting and Rendering, 2nd Edition, 2006, New Riders, Berkeley, CA.

[BRO200701] K. Brown, The use of integral imaging to realise 3D images, in true space, MIRIAD (Manchester Institute for Research & Innovation in Art & Design), Manchester Metropolitan University, UK, 2007.

[CAL193701] C. Calov, Über die Ursache der Verschmelzungsstörung bei der Überschreitung der stereoskopischen Tiefenzone. Das Raumbild Nr. 4 (1937). (Reprint in: Deutsche Gesellschaft für Stereoskopie e.V. Stereo-Report Nr. 31, Berlin, 1982), 1937.

[COB200701] Cobalt Entertainment [Online], 2007. http://www.cobalt3d.com

[COS200701] O.S. Cossairt, J. Napoli, S.L. Hill, R.K. Dorval, and G.E. Favalora, Occlusion-capable multiview volumetric three-dimensional display, *Appl. Optics*, 46, 8, 1244 (10 March 2007).

[CUT199701] J.E. Cutting, How the eye measures reality and virtual reality, *Behavior Research Methods, Instruments, and Computers*, 29, 1, 27–36, February 1997.

[DAC199801] A. Dumbreck, T. Alpert, B. Choquet, C.W. Smith, J. Fournier, and P.M. Scheiwiller, Stereo camera human factors specification, *DISTIMA Technical Report D15*, CEC-RACE-DISTIMA-R2045, 1998.

[GMM199101] L. Gooding, M.E. Miller, J. Moore, and S. Kim, The effect of viewing distance and disparity on perceived depth, in *Proceedings of SPIE Stereoscopic Displays and Applications II*, 259–266, San Jose, CA, August 1991.

[HIN198501] S.P. Hines, Camera Assembly for Three-Dimensional Photography, U.S. Patent 4,557,570, December 1985.

[HOR200901] A.M. Horwood and P.M. Riddell, A novel experimental method for measuring vergence and accommodation responses to the main near visual cues in typical and atypical groups, *Strabismus*, 17, 1, 9–15, 2009.

[IMA200701] IMAX (Online), 2007, http://www.imax.com

[INO199001] T. Inoue and H. Ohzu, Measurement of the human factors of 3D images on a large screen, in *Proceedings of SPIE Large-Screen Projection Displays II*, 104–107, Santa Clara, CA, February 1990.

[IRF200101] W.A. IJsselsteijn, H. de Ridder, J. Freeman, S.E. Avons, and D. Bouwhuis, Effects of stereoscopic presentation, image motion, and screen size on subjective and objective corroborative measures of presence, *Presence*, 10, 3, June 2001.

[IRV200001] W.A. IJsselsteijn, H. de Ridder, and J. Vliegen, Effects of stereoscopic filming parameters and display duration on the subjective assessment of eye strain, in *Proceedings of SPIE Stereoscopic Displays and Virtual Reality Systems VII*, 12–22, San Jose, CA, April 2000.

[ISM200201] W.A. IJsselsteijn, P.J.H. Seuntiëns, and L.M.J. Meesters, State of the art in human factors and quality issues of stereoscopic broadcast television, *ATTEST Technical Report D1*, IST-2001–34396, August 2002.

[JAV200201] B. Javidi and F. Okano, Eds., *Three-Dimensional Television, Video and Display Technology*, Springer Verlag, New York, 2002.

[JLH200101] G. Jones, D. Lee, N. Holliman, and D. Ezra, Controlling perceived depth in stereoscopic images, in *Proceedings of SPIE Stereoscopic Displays and Virtual Reality Systems VIII*, San Jose, CA, January 2001.

[KAU200801] P. Kauff, M. Müller, et al., ICT-215075 3D4YOU, Deliverable D2.1.2: Requirements on post-production and formats conversion, August 2008.

[KLE199801] R. Kleiser, Directing in 3D, *Director's Guild of America Magazine*, February 1998.

[LIG200701] Lightstorm Entertainment (Online), 2007, http://www.lightstormentertainment.net

[LIP198201] L. Lipton, *Foundations of the Stereoscopic Cinema—A Study in Depth*, Van Nostrand Reinhold, New York, 1982.

[LIP199701] L. Lipton, *StereoGraphics Developers' Handbook*. StereoGraphics Corporation, San Rafael, CA, 1997.

[LUS193101] H. Lüscher, *Stereophotographie*. Berlin, Union Deutsche Verlagsgesellschaft, 1931.

[LUS194401] H. Lüscher, Stereoskopische Tiefenzone—Akkommodation und Konvergenz [Stereoscopic depth zone: Accomodation and convergence], Photo-Industrie und Handel, Nr. 7/8 (1944). (Reprint in: Deutsche Gesellschaft für Stereoskopie e.V. Stereo-Report Nr. 31, Berlin, 1982), 1944.

[MIS200401] L.M.J. Meesters, W.A. IJsselsteijn, and P.J.H. Seuntiëns, A survey of perceptual evaluations and requirements of three-dimensional TV, *IEEE Transactions on Circuits and Systems for Video Technology*, 14, 3, 381, March 2004.

[NDB200001] J. Norman, T. Dawson, and A. Butler, The effects of age upon the perception of depth and 3D shape from differential motion and binocular disparity, *Perception*, 29, 11, 1335, November 2000.

[PAC200701] Pace Technologies [Online], 2007, http://www.pacetech.com.

[PAS199101] S. Pastoor, 3D-television: A survey of recent research results on subjective requirements, *Signal Processing: Image Communication*, 4, 1, 21–32, November 1991.

[PAS200201] S. Pastoor, 3D displays: Methods and state of the technology, *Handbuch der Telekommunikation* [Telecommunication Handbook], Köln, Germany, Deutscher Wirtschaftsdienst, 2002.

[RAM194701] F.A. Ramsdell, Apparatus for Making Stereo-Pictures, U.S. Patent 2,413,996, January 1947.

[SMD198801] C.W. Smith and A.A. Dumbreck, 3DTV: The practical requirements, *Television: J.R. Television Soc.*, 9–15, January 1988.

[SUR200501] R. Suryakumar, Study of the Dynamic Interactions between Vergence and Accommodation, Ph.D. thesis, Waterloo, Ontario, Canada, 2005.

[WDK199301] A. Woods, T. Docherty, and R. Koch, Image distortions in stereoscopic video systems, in *Proceedings of SPIE Stereoscopic Displays and Applications IV*, 36–48, San Jose, CA, February 1993.

[YOO200601] H. Yamanoue, M. Okui, and F. Okano, Geometrical analysis of puppet-theater and cardboard effects in stereoscopic HDTV images, *IEEE Trans. Circuits Syst. Video Technol.*, 16, 6, 744–752, June 2006.

[YYS199001] Y.-Y. Yeh and L.D. Silverstein, Limits of fusion and depth judgment in stereoscopic color pictures, *Hum. Factors*, 32, 1, 45–60, 1990.

3

APPLICATION OF VISUAL SCIENCE FUNDAMENTALS TO 3DTV

This chapter takes the concepts of visual science discussed in the previous chapter and applies them to the requirements of supporting 3DTV technology and services. The topics of video capture, stereoscopic projection, polarization, autostereoscopic systems, lenticular lenses, multiviewpoint systems, and holographic displays are covered, among others.

3.1 Application of the Science to 3D Projection/3DTV

Today's TV and most movie display technologies are monocular: Both eyes see the same thing. In contrast, a stereo display presents distinct viewpoints so an observer's left and right eyes see different perspectives, delivering a sense of 3D. As already noted in the past two chapters, stereoscopic 3D video is primarily based on the binocular nature of human perception. 3D video is relatively easy to realize but not totally trivial at the artistic level. Two simultaneous conventional 2D video streams are produced by a pair of cameras mimicking the two human eyes that see the environment from two slightly different angles (see Figure 3.1 for an illustrative example). At the display level, one of these streams is shown to the left eye and the other one to the right eye.

Common means of separating the right-eye and left-eye views include glasses with colored transparencies, polarization filters, and shutter glasses. In the filter-based approach, complementary filters are placed jointly over two overlapping projectors and over the two corresponding eyes (that is, anaglyph, linear or circular polarization, or the narrow-pass filtering of Infitec) [BAK200901]. Although the technology is relatively simple, the requirement to wear glasses while viewing has often been considered as an obstacle to wide acceptance of 3DTV. Also, there are some limitations to the approach, such as the need to retain a head orientation that works properly with the polarized light

Figure 3.1 (Please see color insert following page 160)   Stereoscopic capture of scene to achieve 3D when scene is seen with appropriate display system. Note: In this figure the separation between the two images is exaggerated for pedagogical reasons.

(for example, do not bend the head 45 degrees side to side), the need to be within a certain view angle, and light-intensity issues. Perhaps more importantly, within minutes after the onset of viewing, stereoscopy frequently causes eye fatigue and, in some, feelings similar to that experienced during motion sickness [ONU200801]. There are a number of other mechanisms to deliver binocular stereo, including barrier filters over liquid crystal displays (LCDs) (vertical bars act as a fence, channeling data in specific directions for the eyes), which we discuss later.

As discussed in Appendix 2A, simple planar 3D films are made by recording separate images for the left eye and the right eye from two cameras that are spaced a certain distance apart. The spacing chosen affects the disparity between the left-eye and the right-eye pictures, and thus the viewer's sense of depth. While this technique achieves depth perception, it often results in eye fatigue after watching such programming for a certain amount of time. Nevertheless, the technique is widely used for stereoscopic photography and moviemaking, and it has been tested many times for television [DOS200801].

3.1.1 Common Video Treatment Approaches

In our discussion so far we have indicated that there are several approaches to the treatment of the video images after their immediate capture by a stereo (or multiview) set of cameras. The most common approaches are conventional stereo video (CSV), video plus depth (V+D), multiview video plus depth (MV+D), and layered depth video (LDV). See Table 3.1 [3DT200801], [3DP200801], [3DP200802]. The CSV approach is the mostly likely to see early deployment in commercial 3DTV systems (this topic is revisited in Chapter 4).

There are a number of techniques to allow each eye to view the separate pictures, as summarized in Table 3.2 (based on reference [DOS200801] and described in more details in other sections of this text). All of these techniques work in some manner, but all suffer from (among other shortcomings) the same fundamental psycho-physical problem of accommodation conflict.

3.1.2 Projections Methods for Presenting Stereopairs

We next describe in some detail the principles involved in present-ing—projecting—stereopairs. These principles apply to front- and/or rear-projection TVs (the same general principles and concepts apply to generic TV screens but with appropriate considerations for the medium). Autostereographic means are discussed later.

Stereo projection can be done using two projectors, one for the left-eye image and one for the right-eye image. The lens of each projector is covered with a linear polarizing filter that is at right angles to the other lens/filter (circular polarizers can also be used). The glasses have similarly orientated polarized filters, the net result being that light from the left projector that may be orientated by the polarized filter at +45 degrees reaches the left eye, which has the polarized filter also at +45 degrees, but not the right eye because the right-eye polarized filter is at −45 degrees. A similar situation applies for the right eye: it only receives an image from the right-eye projector [BOU200201]. See Figure 3.2.

There are two ways of creating stereopairs: off-axis projection and toe-in projection. Both approaches work, but the off-axis method is better; the toe-in method is easier to implement but it induces more eye-strain. Make note first that the frustum is the rectangular wedge that one gets if a line is drawn from the eye to each corner of the projection plane—for example, the screen. Figure 3.3 illustrates the two methods (in the figure, the view is looking down on the camera frustum from

Table 3.1 Common Video Treatment Approaches

VIDEO TREATMENT APPROACH	DESCRIPTION
Conventional Stereo Video (CSV)	Conventional stereo video is the most well known and, in a way, the most simple type of 3D video representation. Only color pixel video data are involved, that is, captured by at least two cameras. The resulting video signals may undergo some processing steps like normalization, color correction, rectification, etc., but in contrast to other 3D video formats no scene geometry information is involved. The video signals are meant to be directly displayed using a 3D display system, though some video processing might also be involved before display.
Video plus Depth (V+D)	The video plus depth (V+D) representation consists of a video signal and a per-pixel depth map. (This is also called *2D-plus-depth* by some and *Color plus depth* by others). Per-pixel depth data is usually generated from calibrated stereo or multiview video by depth estimation and can be regarded as a monochromatic, luminance-only video signal. The depth range is restricted to a range in between two extremes Z_{near} and Z_{far}, indicating the minimum and maximum distance of the corresponding 3D point from the camera respectively. Typically, the depth range is quantized with 8 bits, associating the closest point with the value 255 and the most distant point with the value 0. With that, the depth map is specified as a gray-scale image that can be fed into the luminance channel of a video signal and then be processed by any state-of-the-art video codec. For displaying V+D at the decoder, a stereopair can be rendered from the video and depth information by 3D warping with camera geometry information.
Multiview Video plus Depth (MV+D)	Advanced 3D video applications are wide-range multiview autostereoscopic displays and free-viewpoint video, where the user can choose his or her own viewpoint, and require a 3D video format that allows rendering a continuum of output views or a very large number of different output views at the decoder. Today multiview video does not support a continuum of views (as would be the case in holography) but just a discrete set of views. Multiview coding becomes increasingly inefficient for a large number of views. V+D supports only a very limited continuum around the available original view, since view synthesis artifacts generated by the new-view-synthesis algorithms increase dramatically with the distance of the virtual viewpoint. Therefore, a MV+D representation is defined for advanced 3D video applications. MV+D involves a number of complex and error-prone processing steps. Depth has to be estimated for the *N* views at the sender. *N* color with *N* depth videos have to be encoded and transmitted. At the receiver end, the data have to be decoded and the virtual views have to be rendered. The Multiview Video Coding (MVC) standard developed by MPEG supports this format and is capable of exploiting the correlation between the multiple views that are required to represent 3D video.
Layered Depth Video (LDV)	Layered depth video is a derivative and alternative to MV+D. It uses one color video with associated depth map and a background layer with associated depth map. The background layer includes image content that is covered by foreground objects in the main layer. LDV might be more efficient than MV+D because less data have to be transmitted. On the other hand additional error-prone vision tasks are included that operate on partially unreliable depth data that may increase artifacts.

Source: 3DPhone Document All 3D Imaging Phone, 7th Framework Programme.

Table 3.2   Current Techniques to Allow Each Eye to View Distinct Pictures Streams

With appliances (glasses)	Orthogonal polarization	Uses orthogonal (different) polarization planes for each, with matching viewer glasses for each of the left- and right-eye pictures. Light from each picture is filtered such that only one plane for the light wave is available. This is easy to arrange in a cinema but more difficult to arrange in a television display. Test television systems have been developed on the basis of this method, either using two projection devices projecting onto the same surface (especially for rear-projection systems) or two displays orthogonally placed so that a combined image can be seen using a semi-silvered mirror. In either case, these devices are "nonstandard" television receivers. Of the systems with glasses, this is considered the "best."
	Colorimetric arrangements	One approach is to use different colorimetric arrangements (anaglyth) for each of the two pictures, coupled with glasses that filter appropriately. A second is a relatively new notch filter color separation technique that can be used in projection systems (advanced by Dolby).
	Time multiplexing of the display	Sometimes also called "interlaced stereo." Content shown with consecutive left and right signals and shuttered glasses. This technology is applicable to 3DTV. This technique is still used for movie theaters today, such as IMAX, and sometimes used in conjunction with polarization plane separation. In a cathode ray tube (CRT) environment, a major shortcoming of the interlaced stereo was image flicker, since each eye would see only 25 or 30 images per second, rather than 50 or 60. To overcome this, the display rate could be doubled to 100 or 120 Hz to allow flicker-free reception.
	"Virtual reality" headset	Technique using immersion headgear/glasses often used for video games.
Without appliances	Lenticular	This technique arranges for each eye's view to be directed toward separate picture elements by lenses. This is done by fronting the screen with a ribbed (lenticular) surface.
	Barrier	This technique arranges for the screen to be fronted with barrier slots that perform a similar function. Two views (left and right) or more than two (multicamera 3D) can be used. However, since each of the picture elements (stripes or points) have to be laid next to each other, the number of views has an impact on the resolution available. There is a tradeoff between resolution and ease of viewing. Arrangements can be made with this type of system to track head or eye movements and thus change the barrier position, giving the viewer more freedom of head movement.

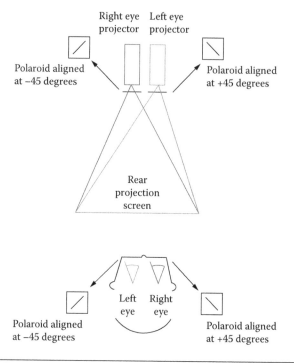

Figure 3.2 Polarized projection.

directly above—there is no vertical parallax and the diagram does not consider the case of a vertical off-axis frustum since the assumption is that the viewer is located centrally to the projection plane, namely the screen). In stereographics the frustums that need to be created are shown in red and blue. The left-hand side (LHS) of the figure depicts the off-axis projection and the right-hand side (RHS) depicts the toe-in projection. Note that the left and right cameras are parallel to each other for the off-axis projection case. The toe-in method rotates the camera frustum so that the two camera direction vectors meet at the focal length. A symmetric frustum as shown in the LHS of the figure is preferable. The way to identify whether or not a stereopair has been created this way is to look for vertical parallax toward the corners of the image: a correctly rendered stereopair will not have any vertical parallax but a toe-in stereopair will have increasing amounts of vertical shift as one moves toward the corners [BOU200201].

The toe-in method is suboptimal because there are problems that originate with positive and negative parallaxes (defined in Appendix 2A). At the optics level, the projection plane is a set distance from the

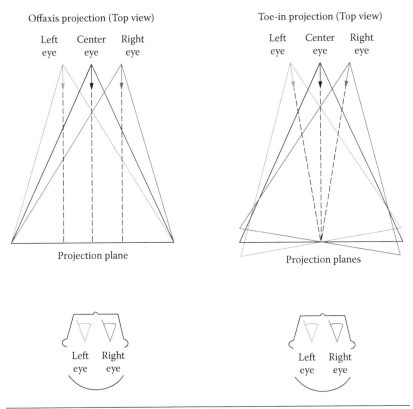

Figure 3.3 Projection methods.

projector that is often called the focal length (but that is better described as the distance for zero parallax). The position on the projection plan of a point in the scene is given by the intersection on the projection plane of a line that passes from the eye to the point in question [BOU200201].

- The case where the object is behind the projection plane is illustrated in the LHS of Figure 3.4. The projection for the left eye is on the left and the projection for the right eye is on the right, and the distance between the left- and right-eye projections is called the horizontal parallax. Since the projections are on the same side as the respective eyes, it is called a positive parallax. Note that the maximum positive parallax occurs when the object is at infinity; at this point the horizontal parallax is equal to the interocular distance.
- If an object is located in front of the projection plane (center portion of Figure 3.4), then the projection for the left eye is on

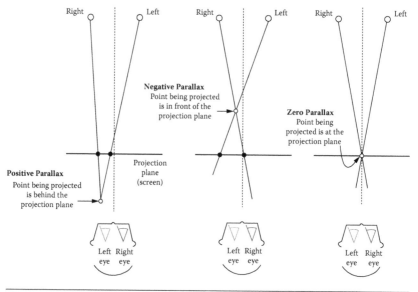

Figure 3.4 Parallax issues in toe-in projection.

the right and the projection for the right eye is on the left. This is known as negative horizontal parallax. Note that a negative horizontal parallax equal to the interocular distance occurs when the object is halfway between the projection plane and the center of the eyes. As the object moves closer to the viewer, the negative horizontal parallax increases to infinity.

- If an object lies at the projection plane, then its projection onto the focal plane is coincident for both the left and right eyes, hence zero parallax (RHS of Figure 3.4).

Hence if on object rendered onto a stereopair has positive parallax, then it will appear behind the screen it is projected onto. If the object appears on the stereopairs in negative parallax, then it will appear to be in front of the screen. Objects that appear in the stereopairs at the same place have zero parallax and will appear to be at the same depth as the screen.

Some researchers make the point that for natural stereoscopic vision, shooting content simply by using two cameras in parallel is not sufficient; parallelization of stereoscopic video is required. These researchers have proposed a system that uses 25 cameras aligned in parallel and/or 100 cameras aligned in line to support this kind of content capture [FUK200901].

3.1.3 Polarization, Synchronization, and Colorimetrics

At the practical level, stereoscopic viewing can be achieved with either *polarized glasses* or *shutter glasses*.

A light wave is an electromagnetic wave that travels through a medium; light waves are produced by vibrating electric charges. A light wave that vibrates in more than one plane is referred to as unpolarized light (for example, light emitted by the sun, by a lamp). Polarized light waves are light waves where the vibrations occur in a single plane. It is possible to transform unpolarized light into polarized light; the process of transforming unpolarized light into polarized light is known as polarization. The most common method of polarization involves the use of a polarized filter. Polarized filters are made of a special material that is capable of blocking one of the two planes of vibration of an electromagnetic wave. When unpolarized light is transmitted through a polarized filter, it emerges with one-half the intensity and with vibrations in a single plane; it emerges as polarized light. A polarized filter is able to polarize light because of the chemical composition of the filter material. The alignment of these molecules gives the filter a polarization axis (see Figure 3.5).

Polarized glasses are similar to regular sunglasses. They are the most popular types of 3D glasses and have been used as a medium for 3D stereoscopic viewing for a long time. They are typically used by large cinema houses and IMAX. Polarized glasses employ the lenses to show

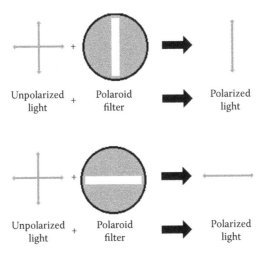

Figure 3.5   Polarization.

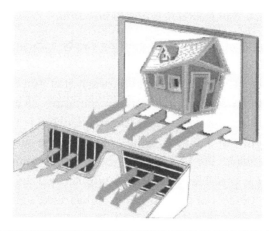

Figure 3.6   Polarized content and use of polarized glasses.

different images to each eye, making the brain construct a 3D image for the viewer. The content (movie) will need to be shot using two cameras (or a single camera with two lenses). Two projectors (left and right), both fitted with polarizing filters on their lenses, then simultaneously show the movie on the same screen. The polarizing filter orients images from the left projector to one plane (say, vertical); and the filter on the right lens orients its images to the plane that is perpendicular to the left one (horizontal). The viewer uses the special glasses that are equipped with differently polarized lenses. The left lens of the glasses is aligned with the same plane (vertical or perhaps 135 degrees) that the left projector is illuminating and displaying the images, and the right lens is aligned perpendicularly to correspond with the plane of the right projector (horizontal, or perhaps 45 degrees). In this fashion, the viewer's left eye sees only the images that the left projector is screening, while the viewer's right eye sees only the images that the right projector is screening (see Figure 3.6). As both the images are taken from different angles, the viewer's brain combines the two to come up with a single 3D image [PAT200901]. There are some drawbacks with this approach, including:

- The amount of light reaching the eyes with polarized glasses is significantly less than without glasses, making the image appear darker than it is;
- There are some limitations such as the need to retain a head orientation that works properly with the polarized light and the need to be within a reasonable viewing angle.

"Shutter glasses" are glasses that alternately shut off the left eye and then the right eye, while the TV emits separate images meant for each eye, thus creating a 3D image in the viewer's mind. Synchronization is key here; shutter glasses allow synchronizing production of left–right images with their alternate acquisition by the viewer's eyes. This is the 3D technology that Panasonic, Sony, and Nvidia are most optimistic for in the near future. The video signal of the TV stores an image meant for the left eye on its even field and an image meant for the right eye on its odd field. The shutter glasses are synchronized with TV via infrared or radio frequency (RF) technology. The shutter glasses contain liquid crystal and a polarizing filter. Upon receiving the appropriate synchronization signal from the TV, the shutter glass is automatically applied with a slight current that makes it dark, as if a shutter were drawn (hence the name). As a result only one eye—say, the left eye—is seeing the image intended for that eye (the left eye). The system perfectly draws the shutters over either eye to make the left eye see the image meant for it on the even field and to make the right eye see the odd field of the video signal. By viewing these two images from different orientations, a 3D image is built up by the viewer's brain [PAT200901]. A limitation of this technology arises from the rapid shuttering, which implies that less light reaches the eye and thus makes the image seem darker than it is or should be.

The most common types of glasses are simple anaglyph glasses. Color encoding (anaglyph) (also known as colorimetric methods) has been used for 3D over the years. The left eye and right eye images are color encoded to derive a single merged (overlapped) frame. Red/blue, red/cyan, green/magenta, or blue/yellow color coding can be used, with the first two being the most common; orange/blue anaglyph techniques are claimed by some to provide good quality, but there is a continuum of combinations. In theaters the content is projected with two color layers superimposed onto one another; when using the tinted glasses, each eye sees a separate visual—say, the red-tinted image through one eye and the blue-tinted one through the other—and the visual cortex combines the views to create the representation of 3D objects. Early anaglyph imaging suffered from a number of issues: The color separation on film was limited, and thus it was difficult to perceive details in 3D scenes; another problem was ghosting, which happens when the image that should be appearing in the left eye also creeps over to the right. Theaters projecting 3D movies with the anaglyph method typically had to install silver screens because the more reflective screen helped keep the two different light signals separated. Anaglyph imaging has improved in recent years; glasses now are typically red and cyan, which result in more realistic color perception [BCH200901].

The largest 3D glasses order ever was reportedly filled by APO Corp. in 2009, with the production of 130 million pairs for the Super Bowl ads. The consensus is that it was not highly successful as 3D but much of the material was unsuitable due to its color (that is, all-white background with people in white suits), or to the fact that it was animated (such as the *Monsters vs. Aliens* trailer). Animations work less well with any stereo method due to their lack of all the rich stereo cues in real-world video. On the other hand, since they are entirely computer generated, changing the colors and brightness of every object to optimize the anaglyph will be much easier than for live action, where the objects will have to be first identified by hand or an image segmentation program [STA200901].

All anaglyphs force one eye to focus at a different emanation plane than the other (the basis of chromostereopsis defined earlier, and the recent ChromaDepth method*); furthermore, the different light levels tend to make one pupil dilate more than the other, and these conditions

---

* The ChromaDepth 3D process enables the creation of "normal" looking color images that can be viewed as two-dimensional images without ChromaDepth 3D glasses, but which appear as 3D when viewed through the ChromaDepth 3D glasses. The ChromaDepth 3D process is entirely new. The ChromaDepth 3D glasses are completely clear, unlike the anaglyph 3D glasses. Although they look simply like clear plastic, each of the lenses in the ChromaDepth 3D glasses actually incorporates a high-precision system of micro-optics that creates a stereopair from a single image. The lenses accomplish this by shifting the image colors in different directions for each eye. The ChromaDepth glasses create quality 3D images from normal 2D images by pulling forward the color red to the foreground and sorting the remaining colors according to their position in the rainbow. The concept of ChromaDepth 3D is straightforward; encode depth into an image by means of color, then decode the color by means of optics, producing a true, stereoscopic, three-dimensional image. Since color is used to represent depth information, the depth-encoded image is a single image. Depending on the subject matter and the skill of the image creator, the effect of the ChromaDepth 3D depth encoding may be to make the image look completely artificial, like a fantasy computer graphic, or to look completely natural. There is a wide range of variations possible with ChromaDepth 3D. The glasses optics come in four different varieties and the depth-to-color relationship (the ChromaDepth 3D mapping) depends on the choice of background color. One can assume that the ChromaDepth 3D glasses are red proud (meaning that red is the foreground color when placed on a black background) and that a black or dark blue background will be used. This combination leads to the most natural looking images, and it has the most consistent set of design rules [CHR200901].

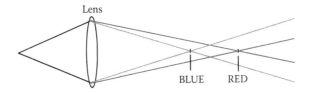

Lens

BLUE   RED

RED light does not focus to the same point as BLUE.
One needs to match the focus with a diopter correcting
the RED light to match the focus point of the other
lens. It can be 5 times sharper with correction!

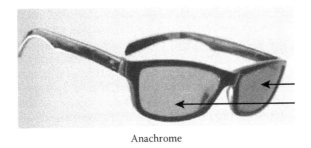

Anachrome

Figure 3.7   Anaglyph glasses.

produce degradation in view. Attempts have been made to ameliorate this situation with glasses that contain a low diopter in one eye; diopter correction addresses the innate focusing problem caused by red filters set close to the eye. Simple optics show that the three colors of red, green, and blue fall into focus at different points on the retina when one is watching a close 3D display, such as a computer or TV monitor. Hence the red-filtered eye gets a much softer image than the cyan, which passes most of the color information—this gives rise to the need to provide correction. Paper anaglyph glasses make sense when distributed in magazines. For 3D images displayed on screens, break-resistant, cleanable, plastic anaglyph glasses are better [ANA200901]. See Figure 3.7.

Theaters showing 3D movies currently use light polarization methods. Table 3.3 provides a quick overview of the two most common methods of theater projection based on reference [BCH200901]. In fact, there are at least eight distinct types of large cinematic screen 3D projection currently available. All techniques that use sheets of plastic polarizer in the projection path have the limitation that these absorb much of the light and so high-brightness projectors will require

Table 3.3 The Two Most Common Methods Used In Theaters

RealD Cinema	Currently the most-widely used 3D movie system in theaters uses circular polarization—produced by a filter in front of the projector—to beam the film onto a silver screen. The filter converts linearly polarized light into circularly polarized light. When the vertical and horizontal parts of the picture are projected onto the silver screen, the filter slows down the vertical component. This effectively makes the light appear to rotate, and it allows the viewer to more naturally move his or her head without losing perception of the 3D image. Circular polarization also eliminates the need for two projectors shooting out images in separate colors. The silver screen, in this case, helps preserve the polarization of the image.
Dolby's 3D system	Used for some *Avatar* screenings. Dolby's 3D system makes use of an exclusive filtering wheel installed inside the projector in front of a 6.5-kilowatt bulb. Effectively, it operates as a notch filter (a notch filter is a band-rejection filter that produces a sharp notch in the frequency-response curve of a system; a filter rejects/attenuates one frequency band and passes both a lower and a higher frequency band). The wheel is divided into two parts, each one filtering the projector light into different wavelengths for red, green, and blue. The wheel spins rapidly—about three times per frame—so it does not produce a seizure-inducing effect. The glasses that the viewer wear contains passive lenses that only allow light waves aligned in a certain direction to pass through, separating the red, green, and blue wavelengths for each eye. The advantage of Dolby's 3D system is that there is no need for a silver screen and it also provides a bright picture, necessary for 3D viewing. Furthermore, a mechanism can be adjusted inside the projector to change the projection method from reflection to refraction—meaning theaters can switch between projecting regular movies and 3D movies. However, the glasses are relatively expensive at around $30 apiece; they are designed to be washed and reused. Altogether, a Dolby 3D projection system costs theaters about $30,000, not including the eyewear.

cooling and degrade the polarizer. Some have dealt with this and other limitations by specifying wire grid polarizers. Conventional thin-film transistors (TFTs) have advantages over current liquid crystal on silicon (LCoS), and so Kodak and others are developing ways to make more complex projectors to enable their use for 3D. The dominant stereoscopic projection system at the moment (marketed by RealD and several others) uses electro-optic switching of circular polarization (CP) with a specially constructed multilayer liquid crystal (LC) plate in front of the projector lens with a silver (that is, aluminized) screen and passive paper or plastic CP glasses for viewing. This is an old idea and goes back at least to the 1940s when the first sheet polarizers were invented; CP at that time was done via a rotating polarized disc (a system resuscitated and now being marketed to the 3D movie industry). Kerr cells (electrically switchable polarizing liquids, in principle identical to the CP switching of LCDs due to the same electrically controlled birefringence) were invented and patented for this purpose about the same time. The achromatic (color-correcting) properties of triple sets of mutually orthogonal half-wave retarders, discovered in 1952, has been researched frequently and most vigorously recently by ColorLink (now part of RealD), and its components and related or alternative display tech is coming in from all the big companies as well as countless smaller ones, such as DigiLens (now owned by SBG Labs), that is, switchable Bragg gratings, LC tech from Rolic, tunable electrowettable diffraction filters from Nokia, and many others [STA200801] (the interested reader may find additional details on this topic in this reference). Stereoscopic projection technology is only marginally more expensive than standard digital projection systems.

Table 3.4 defines for reference some key 3D projection technology terms, based on reference [SUN200901].

The production pipeline for 2D television has developed into a mature and well-understood process over many years; scenes are recorded with cameras from single-view points and then captured image streams are postprocessed, transferred to receivers, and displayed on planar screens. In contrast, the production process for 3D television requires a fundamental rethinking of the underlying technology. Scenes have to be recorded with multiple imaging devices that *may* be augmented with additional sensor technology to capture the three-dimensional nature of real scenes. In addition, the data format

Table 3.4   Some Key 3D Projection Technology Terms

Alternate frame 3D	A method of 3D film stereoscopy involving flashing the right-eye image while the left eye is covered and vice versa; requires the use of synchronized shutter glasses; contrast with polarized light method
Anaglyph	A primitive and inexpensive red/cyan method of 3D transmission associated with the colored glasses
Anamorphic lens	A lens used to change the shape and perspective of the frames as printed on film; used to create super-wide projection
Autostereograms	3D images that can be viewed without the use of glasses or any other kind of headgear attached to the viewer; e.g., Philips 3DTV and Magic Eye
CinemaScope	The first anamorphic film format at the cinema; allowed for super-wide screen projection of films; rival to 3D in the Golden Era
Cinerama	Triple film strip, triple screen, super-wide cinema gimmick used to bring the audiences back to the movie theaters; rival to 3D in the 1950s
Digital cinema	Movie theaters that have switched from using traditional 35 mm film projectors to DLP units
DLP	Digital light processing; a Texas Instruments–owned technology used to power non-film projectors both at home and in digital cinemas
Dolby 3D	Lesser-used color 3D cinema system featuring the same principles as anaglyphs but with left- and right-eye images separated with different wavelengths of the same colored light
Front projection	The standard method by which pictures are passed by a bulb and the light throws images focused through a series of lenses onto a screen in front of the audience
Golden Era	The first mass popularization of 3D films at the cinema; took place in the early 1950s
IMAX	Super-large film format traditionally printed on 70 mm stock and projected onto screens in the region of 22x16 m in size; created by IMAX Corporation of Canada and now also in digital format
Infitec glasses	Super-anaglyph glasses used in the Dolby 3D cinema system; they filter discrete wavelengths of all colors
Interlacing	Technique used in alternate frame 3DTV where only every other line of the screen is used to display each eye's image at any one time; saves on processing power but halves resolution
LCD shutter glasses	Liquid crystal glasses that turn opaque when a charge is sent through them; used in alternate frame 3D transmission in both XpanD cinemas and in the home
Lenticular lenses	Curved optics that allow both eyes to see a different image of the same object at exactly the same time; used on holographic stickers
Over and under	The common name of the method of frame placement used in the polarized light-based Space-Vision 3D of the 1960s
Parallax barrier	Device used on the surface of non-glasses-based 3DTV system with slits that only allow the viewer to see certain vertical columns of pixels at any one time
Polarized light	Light filtered such that it only contains waves oscillating in a specific orientation; used in the majority of 3D cinemas
Pulfirch effect	The apparent swinging of a pendulum in elliptical rotation viewed with one eye closed, when the pendulum is actually moving in just one plane; the principles were used to create 3D illusions in an episode of *3rd Rock from the Sun*

Table 3.4 (*Continued*) Some Key 3D Projection Technology Terms

RealD Cinema	3D cinema system adopted by about 80 percent of the 3D movie theaters; uses light polarized in different directions to create the two separate left- and right-eye images. RealD started businesses concerning stereoscopic technologies in industrial and also military purposes 30 years ago. Recently, it advanced into the field of movies and has supported the production of about 35 3D movies
RealityFusion camera system	Digital, high-definition, stereoscopic camera developed by James Cameron and Vince Pace to shoot *Ghosts of the Abyss*
Rear projection	DLP units for the home with a TV-shaped box containing a projector that throws the image a short distance onto the back of the screen; contrast with front projection
S3D gaming	True stereoscopic immersive 3D gaming; contrast with 3D graphics that allow one to examine the depth of an object on screen but in only one 2D plane at any one time
Side by side	Stereographic material where the similar left and right images are presented next to one another; defocusing or crossing eyes is used to create their 3D effect in stills and an anamorphic lens[a] when through a cinema projector
Silver screen	A silver painted screen used to reflect light without degrading any polarization; key in 3D IMAX and other RealD cinema types
Space-Vision	The first single-strip solution to 3D film projection, invented by Arch Oboler in the 1960s; uses an over and under technique of each eye, and involves frame placement of a special lens on the projector that polarizes each one in the opposite direction to the next
Stereoscopy	Any technique used to create the illusion of depth from flat images or video; e.g., anaglyph, Magic Eye, polarizing glasses, etc.
Stereoscope	Any device used to view side-by-side still images for a 3D effect; popular from the 1850s on
Stereovision	Format of 3D film produced in 1970s that ran both left- and right-eye images side by side on the same strip and used an anamorphic lens to widen the squashed frames before being independently polarized and projected onto the same screen
Teleview	The earliest form of alternate frame sequenced 3D stereoscopy invented in 1922; required shutter glasses fixed to cinema seats and was only ever installed in one cinema
XpanD	Lesser-used 3D cinema system involving LCD shutter glasses and rapid alternate frame projection to form the stereoscopic effect

Source: D. Sung, 3D Dictionary, *Gadget News, Reviews, Videos*, http://www.pocket-lint.com, 21 September 2009.

[a] Optical anamorphic lens are used to recreate an original aspect ratio. The most common application is to use the anamorphic lens to convert full 4:3 native format into 16:9 widescreen format. Basically, an anamorphic lens is a lens that optically distorts the image. The anamorphic lens was first developed in the film industry when they wanted to use standard 35 mm film to record images in widescreen format. The way this was done was to fit the film camera with a widescreen format lens that optically compressed the image so that it would fit into a 35 mm film frame. Then when the film was played through a projection system, the projector was fitted with another lens that reversed the distortion. In that way the compressed image that was recorded on the 35 mm film was projected onto the screen in natural, uncompressed widescreen format [POW200301].

used in 3D television is a lot more complex. Rather than normal video streams, time-varying computational models of the recorded scenes are required that comprise descriptions of the scenes' shapes, motions, and multiview appearances. The reconstruction of these models from the multiview sensor data is one of the major challenges that we face today. Finally, the captured scene descriptions have to be shown to the viewer in three dimensions, which requires completely new display technology [3DT200701].

3.2 Autostereoscopic Viewing

There are some limitations of stereoscopic 3D viewing: Glasses may be cumbersome and expensive (especially for a large family), and without the glasses, the 3D content is unusable. Autostereoscopic 3D television eliminates the use of any special accessories (LG and Panasonic, among others, are working on this technology). Autostereo implies that the perception of 3D is in some manner automatic and does not require devices such as glasses—either filtered or shuttered. Autostereoscopic displays use additional optical elements aligned on the surface of the screen to ensure that the observer sees different images with each eye. At the technical level, the term "autostereoscopic" includes holographic, volumetric, and integral imaging; however, it has been used in the 3DTV industry to describe displays such as lenticular or parallax-barrier binocular systems; we follow this convention here. 3D autostereoscopic displays are still in research phase at this time, but some autostereoscopic (multiview) displays are becoming available. As noted in Chapter 1 autostereoscopic technology may be appropriate for mobile 3D phones, and there are several initiatives to explore these applications and this 3D phone-display technology.

While current-generation 3D video applications employ stereoscopic or multiview 3D video by showing slightly shifted views to the right and left eyes with filters based on glasses, autostereoscopic displays that use parallax barriers or lenticular lenses support 3D perception without requiring the user to wear glasses. Typically, autostereoscopic displays present multiple views to the observer, each one seen from a particular viewing angle along the horizontal direction. A typical multiview 3D display device shows eight or nine views simultaneously to allow a limited free-viewing angle [TEK200901].

However, the number of views comes at the expense of a loss of resolution and brightness.

Current limitations of autostereoscopic displays not only impact the "big screen" of a home theater but also the "small screen" of a mobile device. Both resolution and brightness loss are problematic on a small screen, battery-driven mobile device. As mobile devices are normally watched by only one observer, two independent views are sufficient for satisfactory 3D perception. There are only a few vendors currently with announced prototypes of 3D displays targeted for mobile devices; all of them are two-view, TFT-based autostereoscopic displays [3MO200901]. It should be noted that while the bandwidth requirements for real-time 3D video (3DV) transmission service are lower (say, 2x384 kbps or 2x512 kbps with CSV), the wireless bandwidth is at a premium; this premium may be part of the calculus/business case related to establishing a 3DV service for cell phones.

As noted, *lenticular lenses* and *parallax barriers* are the current constituent-enabling technologies. We discuss these next.

3.2.1 Lenticular Lenses

Lenticules are tiny plastic lenses pasted in an array on a transparent sheet that is then applied onto the display surface of the LCD screen (see Figure 3.8). When looking at the cylindrical image on the TV, the left and right eyes see two different 2D images that the brain combines to form one 3D image. The lenslet or lenticular elements are arranged to make parts of an underlying composite image visible only from certain view directions. Typically, a lenticular display multiplexes separate images in cycling columns beneath its elements, making them take on the color of selected pixels beneath when viewed from different directions. LCDs or projection sources can provide the pixels for such displays [BAK200901]. A drawback of the technology is that it requires a very specific "optimal sitting spot" for getting the 3D effect, and shifting a small distance to either side will make the TV's images seem distorted.

3.2.2 Parallax Barriers

The parallax barrier is the more consumer-friendly technology of the two and the only one that allows for regular 2D viewing. This technology is

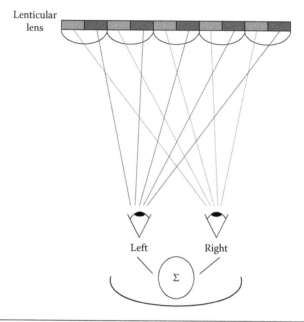

Figure 3.8   Lenticules.

pursued by manufacturers such as Sharp and LG, among others. The parallax barrier is a fine grating of liquid crystal placed in front of the screen, with slits in it that correspond to certain columns of pixels of the TFT screen (see Figure 3.9 based partially on reference [BAK200901]). These positions are carved so as to transmit alternating images to each eye of the viewer, who is again sitting in an optimal "sweet spot." When a voltage is applied to the parallax barrier, it channels direct light from

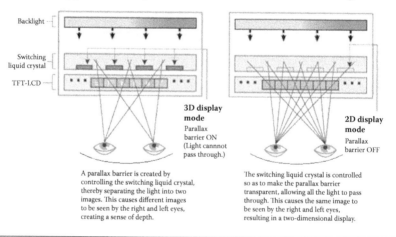

Figure 3.9   Parallax barrier.

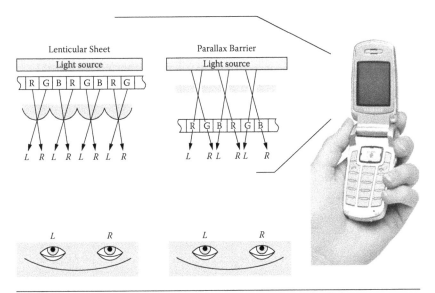

Figure 3.10   Application of autostereoscopic technologies to 3D mobile telephones.

each image slightly differently to the left and right eyes, again creating an illusion of depth and thus a 3D image in the brain [PAT200902]. The parallax barrier can be switched on and off, allowing the screen to be used for 2D or 3D viewing. However, the need still exists to sit in a sweet spot, limiting the usage of this technology.

Autostereoscopic technology will likely not be part of early 3DTV deployments. For example, Philips reportedly folded an effort to define an autostereoscopic technology that does not require glasses because it had a narrow viewing range and had a relatively high loss of resolution and brightness [MER200901].

Figure 3.10 depicts the possible application of these technologies to 3D mobile telephones.

3.3 Other Longer-Term Systems

A number of areas are not covered in this book beyond the mention herewith: multi-viewpoint 3D system, integral imaging,* volumetric displays, and holography. These provide ultimate autostereoscopic viewing.

- **A multi-viewpoint 3D system** is a system that provides a sensation of depth and motion parallax based on the position and

---

* Some additional discussion of integral imaging is provided in Chapter 5.

motion of the viewer. At the presentation (display) side, new images are synthesized based on the actual position of the viewer. The multi-viewpoint system requires a head tracker that measures the 3D viewer position with respect to the stereo display. (For a full multi-viewpoint system that provides any viewpoint, the system needs a head tracker that measures the exact 3D position of the viewer.)

- **Integral imaging (holoscopic imaging)** is a technique that provides autostereoscopic images with full parallax. In the capture subsystem, an array of microlenses generates a collection of 2D elemental images onto a matrix image sensor [such as a charge-coupled device (CCD)]; in the reconstruction/display subsystem, the set of elemental images is displayed in front of a far-end microlens array, providing the viewer with a reconstructed 3D image.

- **Holography** is a technique for generating an image (hologram) that conveys a sense of depth but is not a stereogram in the usual sense of providing fixed binocular parallax information. Holograms appear to float in space and they change perspective as one walks left and right; no special viewers or glasses are necessary. (Note, however, that holograms are monochromatic.) To create a hologram, illumination by laser beams is required and the process is very involved.

- **Volumetric/hybrid holographic** uses geometrical principles of holography in conjunction with other volumetric display technologies.

In addition some researchers have advocated image analysis, where one decouples the image capture and image display elements; in these proposed, longer-term systems, the captured scene is converted to an abstract 3D moving scene; the 3D scene is then rendered at the display end as a function of the display technology employed (there are a number of different ways of rendering 3D info into a consumable form).

3.3.1 Multi-Viewpoint 3D Systems

A new-generation natural 3D display may be required to have the following features [TAK200201]:

- Simultaneous observation by several persons
- No need to wear special 3D glasses
- High-presence provision
- No contradiction to human 3D perception
- Offering color and moving images

Considering the present device technology, the most promising candidate to meet these requirements is a multiview 3D display. There has been much research in building devices to provide autostereoscopic viewing (as mentioned above), but this has been done almost exclusively in the context of presenting computer graphics (CG) models for off-line visualization. Little capability exists for capturing and presenting live multi-viewpoint 3D video. Even a two-view binocular display presents a limited experience—all observers receive the same fixed perspective; viewed statically, a scene takes on a cardboard-cutout appearance since proprioceptive* expectations are not met [BAK200901], and without accounting for changing position, a viewer's motion makes the static views take on the appearance of a skewing/twisting scene. The goal of immersion is to provide a natural, authentic, stable, and compelling experience. These techniques serve to recreate a light field that a user can sample by positioning himself at different locations within its presentation range. The notion of light field comes from the fact that the field of light that would arise from the true scene is recreated and directed as it would be were the scene actually present.

To achieve a multi-viewpoint display system, a number of projectors generate projections that are fully superimposed on the screen, with their sources positioned approximately behind and above each of participants. Such an arrangement of projectors and cameras delivers more natural eye contact and greater gesture awareness, with studies showing a measurable increase in trust of about 30 percent versus standard (that is, non-multiview) video conferencing. See Figure 3.11 for an example from reference [BAK200901]. Free-viewpoint TV (FVT) technologies have been attracting attention as a promising method for viewing sports and performances; FVT records actions

---

* Proprioception is the unconscious perception of movement and spatial orientation arising from stimuli within the body itself.

Figure 3.11 (Please see color insert)   HPLabs-UCBerkeley Multiview Facility: (left) projectors; (right) display and cameras.

and visual scenes with multilens cameras that enable viewing images from any viewpoint [MAS200901].

In a typical multiview system, several horizontal parallax images are displayed into the corresponding horizontal directions. Observers can percept 3D images with the binocular disparity, the vergence, and the motion parallax. The accommodation function, however, does not work in current-generation multiview 3D displays. Some have reported that the accommodation function may work when high-density horizontal parallax images are displayed simultaneously. Because parallax images are displayed with a very small angle interval, rays actually converge in 3D space and more than one parallax image enters into the pupil of an observer's eye. When the eye focuses on the plane where a 3D object is displayed, as shown in Figure 3.12(a), parallax images are superimposed on one another to form a sharp image on the retina. Otherwise, a blurred image is formed as shown in Figure 3.12(b). Observers are thus released from the fatigue that is the fatal problem of the conventional 3D displays, caused by the conflict between the accommodation and the vergence [TAK200201].

While this technology is interesting (for theatrical and industrial applications), it is not being considered at this time for short-term introduction into the home.

3.3.2 Integral Imaging/Holoscopic Imaging

"Integral imaging" is yet another technology for 3D imaging. This is similar, if not nearly identical, to the concept of holoscopic imaging discussed below.

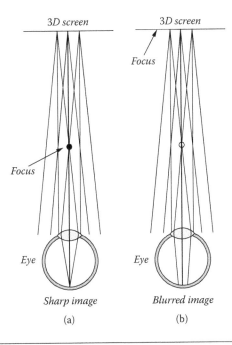

Figure 3.12 Accommodation and high-density horizontal parallax images. (a) eye focuses on a 3D object, and (b) eye focuses on the 3D screen.

3D image-display systems give a sense of improved realism over conventional 2D systems, by offering both psychological and physiological depth cues to the human visual system. The addition of parallax information creates binocular disparity, giving rise to depth perception. Most 3D display systems use stereoscopic techniques to stimulate binocular disparity and binocular convergence. Stereoscopic systems, however, do not always offer comfortable or natural viewing; additional cues such as motion parallax and accommodation should also be exploited. Integral imaging is a 3D display technique that overcomes the limitations of purely stereoscopic systems by including these additional cues [BRO200701]. Integral imaging was first proposed by M.G. Lippmann in 1908 and has been further developed by many others since that time. Integral imaging is seen as an attractive autostereoscopic technique.

Integral imaging allows the acquisition and autostereoscopic display of 3D scenes with full parallax and without requiring the use of any additional viewing devices like special glasses. Integral imaging is based in the intersection of ray cones emitted by a collection of 2D elemental images that store the 3D information of the scene [MAR200801]. The basic concept of integral imaging is to capture many 2D pictures of an object simultaneously from different angles, and then optically projecting the pictures back to the geometric location of the object to re-create the 3D image. Lenslet arrays (micro-lens arrays) are generally used in both capture and reconstruction. Extension of the technique for motion pictures and TV is possible [ONU200601].

The autostereoscopic capability is achieved by taking a set of images of the object in question using an array of microlenses (similar but not identical to a lenticular lens) where each individual lens looks at a different viewing angle. The system generates a number of closely packed, distinct microimages that are viewed by an observer through an array of spherical convex lenses. This approach reproduces a light field that re-creates stereo images that exhibit parallax when the viewer moves. See Figure 3.13; observe in this picture that although all the elemental images are imaged by the corresponding microlenses onto the same plane (known as the reference image plane), the 3D

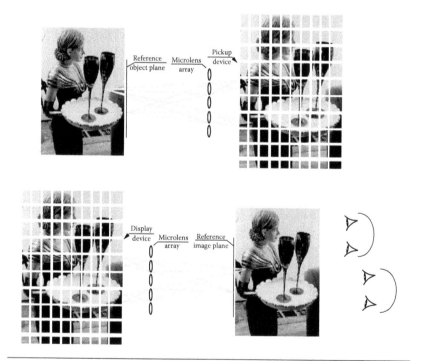

Figure 3.13 (Please see color insert)   Integral imaging principles.

scene is reconstructed in the image space by the intersection of the ray bundles emanating from each of the microlenses.

Notwithstanding recent progress, integral imaging systems still pose problems in their basic configuration. The limitation of viewing angle is one of the main optically related issues. In general, the viewing angle in integral imaging depends on the physical properties of the elemental lens in an array. To overcome this problem, the idea of viewing-angle-enhanced integral imaging using lens switching has been proposed, but its real implementation was limited because the lens switching was performed by a mechanical method [JUN200201]. Furthermore the reconstructed images that integral imaging provides are real (in front of the film) but pseudoscopic—that is, reversed in depth. Also, owing to the spread of ray bundles for out-of-focus parts of the scene, the reconstructed images have very poor depth of field (DOF). To attenuate the DOF problem, some researchers have experimented with binary amplitude modulation during the capture stage. An attempt to overcome the problem of pseudoscopic reconstruction was made by some who proposed to record a second set of elemental images using the reconstructed image as the object. When the second set of elemental images is used in a second display stage, a real, undistorted, orthoscopic 3D image emerges. However, the approach does not completely solve the pseudoscopic-to-orthoscopic (PO) conversion problem [MAR200601]. (Orthoscopic means having correct vision or producing it, giving an image in correct and normal proportions with a minimum of distortion, or being free from optical distortion or designed to correct distorted vision. For example, an orthoscopic lens is a lens that is free of spherical aberration and magnifies an image uniformly throughout the field.)

For 3DTV obtained via a transmission network, the key drawback relates to the very large amount of transmitted scene data that this approach entails.

Some have used the term *3D holoscopic imaging*. This is presented in publications as a technique that is capable of creating and encoding a true volume-spatial optical model of the object scene in the form of a planar intensity distribution by using unique optical components [TSE200901], [MON200101], [AGG200901], [AGG200601], [DAV199201], [BOG198901]. "Holoscopic imaging" was coined by proponents to distinguish the principle from both stereophotography and holography; we view it here as basically being similar, if not identical to, integral imaging. The visual clues involved in depth perception are by and large missing from conventional stereo imagery. Holoscopic imaging seeks to produce a real orthoscopic image in full color without the practical difficulties associated with holography or the limitations of traditional lenticular stereophotography. The latter achieves a stereo image by presenting changing sets of stereo pairs as the viewpoint moves around horizontally, but there is no vertical parallax and the scale effect causes images to appear as if cut out of cardboard. In addition, correct perspective can only be observed from a single fixed distance. In order to achieve full vertical and horizontal parallax, it is necessary to use an array consisting of a large number of microlenses rather than cylindrical lenticles. Each microlens forms its own image from its own point of view, with its own unique perspective. When processed to a positive transparency, the film is viewed behind the original array of microlenses and the image is then returned by each lens to the original position of the object. The more lenses that are involved, the better will be the definition, up to the point where diffraction takes control; this limits the ultimate image resolution. In addition, a more serious problem is that when the image is viewed in the normal way from the object side: It is pseudoscopic (inside out). The goal is then to reinvert the perspective and produce an orthoscopic image. Of late the concept has evolved to use a back-to-back microlens configuration to reverse the emergent angle of the rays and to use a high-density microlens array for the image plane and a large-scale microlens array for the field screens, one on each side to improve the depth of the image. In practice such a camera has to have a large diameter (220 mm) in order to capture sufficient parallax, and the array of macrolenses has to be aligned very accurately indeed (individual variations of less than 11 seconds of arc). The result is a full optical image in 3D in space, not a stereogram (see Figure 3.14).

3D holoscopic imaging is similar to holography in that 3D information recorded on a 2D medium can be replayed as a full 3D optical model; however, in contrast to holography, coherent light sources are not required. This conveniently allows more conventional live capture and display procedures to be adopted. A 3D holoscopic image is recorded using a regularly spaced array of small lenslets closely packed together and in contact with a recording device. Each lenslet views

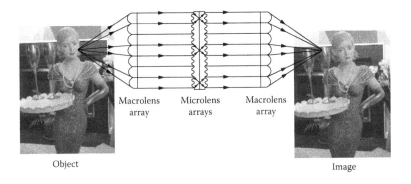

Macrolens array   Microlens arrays   Macrolens array

Object

Image

Figure 3.14 (Please see color insert)   An approach to holoscopic imaging.

the scene at a slightly different angle than its neighbor, and therefore a scene is captured from many viewpoints and parallax information is recorded. It is the integration of the pencil beams that renders 3D holoscopic imaging unique and separates it from Gaussian imaging or holography. A 3D holoscopic image is represented entirely by a planar intensity distribution. A flat-panel display—for example, one using LCD technology—is used to reproduce the captured intensity-modulated image, and a microlens array reintegrates the captured rays to replay the original scene in full color and with continuous parallax in all directions (both horizontal and vertical). With recent progress in the theory and in microlens manufacturing, holoscopic imaging is becoming a practical and prospective 3D display technology and is attracting much interest in the 3D area. It is now being considered as a possible strong candidate for next-generation 3DTV. This 3D holoscopic content will be interactive and expressive, allowing for new visual sensations since it is inherently more interactive than other kinds of video because 3D objects can be extracted from the 3D holoscopic video more easily. This will allow more efficient object segmentation in 3D space to make the objects in the video more "selectable" as 3D holoscopic objects. This new 3D format would revolutionize TV content production and interaction and will lead to new forms of storytelling and content manipulation [TSE200901].

3.3.3 Holographic Approaches

To create content that can be seen in 3D causing no eye fatigue and allowing viewer head movement, one needs to employ holographic methods. Holography is an inherently 3D technique for the capture of real-world objects. Many existing 3D imaging techniques are based on the explicit combination of several 2D perspectives (or light stripes and so forth); the advantage of holograms is that multiple 2D perspectives can be optically combined in parallel in one step independent of the hologram size. Holography is an interferometric technique that allows one to record and reconstruct both the amplitude and the phase of an optical wave front. A hologram records *all* of the information available in a beam of light—not just the amplitude, as in traditional photography, but also the phase of the light [REA200901]. The first off-axis holograms were created in the early 1960s when lasers became available; digital holography techniques followed and eventually holographic cameras appeared. Experimental holographic motion pictures where produced in the late 1980s [ONU200601]. Recent developments in this field suggest that holographic 3DTV displays may become available at some point in this decade. Holography requires large amounts of information to be recorded, stored, transmitted, and displayed, placing a challenge on a 3DTV end-to-end system that would attempt to use this technology at this time.

Consider the way light illuminates objects in real life. When light hits an object, waves of all wavelengths, except that of the "color" of the object, are absorbed by the object surface; the wave with the remaining wavelength of light is reflected and refracted. A viewer's eyes can be in the path of this light and thus be able to "see" the object via the small aperture of the eyes' lenses. However, the light that passes through space toward the viewer's eyes is not a point source of light but it is the summation of light as it emerges from all points and all angles. This totality of light rays is termed the "object wave." The recording of this object wave will produce the complete fatigue-free reproduction of the real image for us to see, if we could capture it. The object wave has magnitude, wavelength, and phase. If one could capture all three, one could record the object wave, and if one has all information about the object wave, issues such as the accommodation conflict disappear. Unfortunately, we do not yet have the means to record more than the amplitude of the wave. The use of holography allows one to record the object wave as an amplitude only; the wavelength recording problem is addressed by illuminating the object with a single wavelength laser beam and the phase recording problem is addressed by folding in the phase information to the amplitude information (this being done by creating an interference pattern with an original laser beam that illuminates the object—the interference pattern created can be recording on a plate). For re-creation, the amplitude and phase information can be restored by illuminating the plate: The outcome is a single wavelength, monochromatic version of the object wave. This is clearly not the ultimate desired outcome but it is a step in the direction of true 3D capture. At present there are no electronic means of recording time-successive slices of component colors of the interference pattern of an object wave, and the storage requirements for holography are seen as being prohibitive [DOS200801]. Advances in nanotechnology and nanophotonics may eventually address some of these issues [MIN200501]. Figure 3.15 depicts the basic principles of holographic image capture and reconstruction.

The current state of the art for television with holographic projection (HoloTV) systems is monochromatic projection of brilliant pictures onto a treated glass screen coated with a semitransparent film, causing viewers to see a picture hanging in midair. In the future, HoloTV

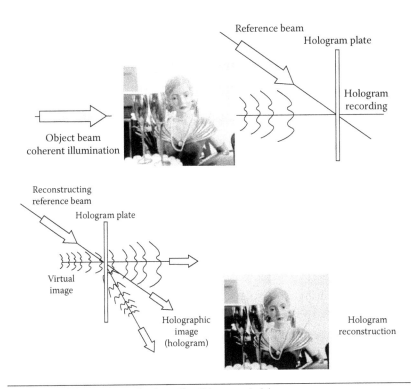

Figure 3.15 (Please see color insert)   Hologram principles.

systems will likely project full 3D images in midair without a projection surface of any kind. Viewer movement may be recognized by a computer via the use of a device worn by a person or through analysis of visual pictures alone by a computer [ONU200801]. Some prognosticate early-stage holographic 3D display should become available by 2015 [HDH200701].

3.3.4 *Volumetric Displays/Hybrid Holographic*

Some high-end systems have appeared that use holographic geometrical principles with special focus on reconstructing the key elements of spatial vision. The "voxel" (a voxel is the 3D analog to a pixel in a 2D image) of the holographic screen emits light beams of different intensity and color to the various directions. A light-emitting surface composed of these voxels acts as a digital window or hologram and will be able to show 3D scenes. Volumetric displays form the image by projection within a volume of space without the use of a laser light reference but have limited resolution. Some call these systems "hologram-like 3D displays."

Volumetric displays are a present-day technology that has seen more broad deployment in high-end scientific/industrial environments. Volumetric display devices are graphical display devices that form a visual representation of an object in 2D, in contrast to the planar image of traditional screens. Volumetric displays create 3D imagery via the emission, scattering, or relaying of illumination from well-defined regions in (x,y,z) space. Each (x,y,z) spatial position (voxel) is displayed physically to form a real-volume 3D model "floating in space." Holographic and high-multiview displays are not always considered to fit this category. The ability to have a large number of simultaneous viewers is unique to true volumetric displays. Stereoscopic displays and other alternative 3D systems often have a limited number of viewable positions, or sweet spots, and limited 3D depth cues. Volumetric displays are finding applications in the fields of air traffic control, situational awareness, medical image viewing, battlefield management, submarine navigation, scientific data visualization, molecular modeling, viewing complex mathematical surfaces, visualization of biological and chemical structures, games, 3DTV, sport stadium display, and education [TEC200901].

Volumetric 3D displays have been around for nearly a century, but they face several challenges that have prevented widespread deployment. Some volumetric displays involve thousands of 3D pixels (or "voxels" for "volume elements") that either absorb or emit light from an "isotropically emissive light device" (IEVD). The voxels are projected onto a screen that rotates, say, 24 times per second, creating a 3D image. Because the image is composed of either the presence or absence of light, it creates an x-ray-like effect of the input data. One of the major issues with volumetric displays is the difficulty in portraying shading of object surfaces and displaying opaque objects, both of which are common in real-world situations. To improve these traits, researchers are developing new techniques that modify the original input data in such a way as to allow the light rays to produce more shading and darkening effects. The shading effects can capture surface curvature and texture to provide improved image quality [ZYG200801]. Unlike holographic techniques, volumetric displays do not require extremely large amounts of computation for data processing.

Desiderata for "true 3D (volumetric) systems" include the following (also see Table 3.5) [HOL201001]:

- No glasses needed; the 3D image can be seen with unassisted naked eye.
- Viewers can walk around the screen in a wide field of view, seeing the objects and shadows moving continuously as in the normal perspective. It is even possible to look behind the objects; hidden details appear while others disappear (motion parallax).
- Unlimited number of viewers can see simultaneously the same 3D scene on the screen, with the possibility of seeing different details.
- Objects appear behind or even in front of the screen, like on holograms.
- No positioning or head tracking applied.
- Spatial points are addressed individually.

Some specialized systems are available for medical applications, scientific imaging, theme parks, air traffic control, simulations, game arcades, and other entertainment business areas (for example, see www.holografika.com; also see Figure 3.16—clearly a system such as the one shown in Figure 3.16 is *not* for home use at this time).

Table 3.5   Desiderata for True 3D Systems

FEATURES	USER BENEFITS
Continuous motion parallax	The pictures don't "jump" between the views, horizontal perspective
There is no contradiction between eye convergence and focusing	No discomfort, seasickness, or disorientation
Pixels (voxels) can be individually addressed	The point of a given view does not move if the viewer is moving and is exactly there where it seems to be
Wide viewing angle, optional viewing distance	Collaborative use
No eye or head tracking used	Not necessary to stand in one point, free motion possible
	No latency
No invalid zones in the field of view	The 3D view can be seen in the entire field of view given
The points can be generated anywhere in the field of view	Optional shaped objects or 3D view can be visualized
	Objects could appear behind and in front of the display screen
Ability to show hidden edges	Wide scale of displayed images, textures vs. wire frame images only
Direction-selective light emittance of each pixel	Windowlike view, all light beams are present as in natural view
Full compatibility with optional software environment	Ability to display any 3D information and to use different 3D software without restrictions
Compatible with current displaying conventions	Easy replacement of desktop 2D monitors
High refreshment rate possible	Motion picture, interactivity
Brightness	Good visibility under various ambient lighting conditions
Resolution (x,y)	Clear, sharp pictures
Depth resolution (z)	Good depth of the 3D view
Color resolution	True colors
Practical display sizes	System can be built from small to large sizes
No moving parts	Reliability, proper lifetime, mobile applications, no noise

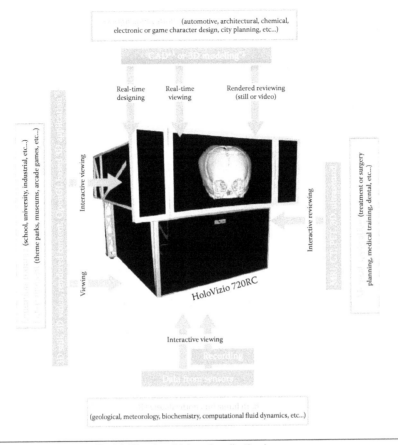

Figure 3.16   Holografika hybrid system (high-end applications).

Table 3.6 lists a few of the volumetric displays available at press time; some of these are products, others are just prototypes. This kind of volumetric display technology will not be employed in 3DTV any time soon.

3.4 Viewer Physiological Issues with 3D Content

This section describes two typical 3D viewer issues* that have to be taken into account by designers of 3DTV:

- The accommodation problem
- Infinity separation

---

* This discussion is based on reference [DOS200801].

Table 3.6   Partial List of Volumetric Displays Available at Press Time

Holografika HoloVizio	HoloVizio is a technology capable of showing natural 3D images available on the market today. Watching 3D images on HoloVizio does not require the viewers to wear glasses or polarized color filters (anaglyph). One simply stands in front of the screen and experiences a high-quality 3D image, created by the latest technology. It is possible to look into and behind objects, hidden details appear, while other parts of the displayed object disappear the same way as it happens at real 3D scenes in life (motion parallax). Unlimited number of viewers can watch simultaneously the same 3D image on a HoloVizio 3D display with the possibility of seeing different details of the scene. Every viewer sees his or her own perspective of the same 3D scene.
USC Lab	Projects a video frame into a rapidly spinning mirror; close to 5,000 individual images are reflected every second within the surface area and come together to create a real-space three-dimensional object. The system uses spinning mirrors, high-speed DLP projections, and very precise math that figures out the correct axial perspective needed for a 360-degree image. Because the images projected from the mirror jump out "toward multiple viewpoints in space," the USC team created a formula that renders individual projections at different heights and traces each projected beam back to the display area to find the correct position of the viewer. The system also updates itself in real time (at 200 Hz), adjusting to the height and distance of the viewer, producing an image that will "stay in place" [FER200801].
Technest's 3D VolumeViewer	The 3D VolumeViewer allows users to view a 3D object from multiple points of view without the use of special glasses and without the perspective problems associated with lenticular approaches. Vendor claims that this true 3D-display system provides both physiological and psychological depth cues required by the human visual system to perceive 3D objects. The system uses state-of-the-art DLP projection technology from Texas Instruments to provide a radiant, high-density picture that is viewable from any angle. The 3D VolumeViewer system provides both physiological and psychological depth cues required by the human visual system to perceive 3D objects.
Perspecta Spatial 3	The Perspecta volumetric display generates 3D images and is based on a sweeping plane that performs 24 rotations per second. The system consists of a 20-inch dome that plugs into a PC to display full-color and full-motion MRI, x-ray, CT, and nuclear medicine images in 3D space.
Heliodisplay	Re-creates 2D projections into floating 3D illusions, but it is not holographic. The Heliodisplay uses patented and proprietary technology to create an almost invisible trilayered, out-of-phase field to generate the surface required to accept projection of video or images into free space. The image is displayed into two-dimensional space (i.e., planar); Heliodisplay images appear 3D, even though in fact the images are planar. This allows for easy display of visual presentation material with a 3D appearance since there is no physical depth reference. Images appear more three-dimensional that 3D displays. Images can be seen up to 75 degrees off aspect for a total viewing area of over 150 degrees, similar to an LCD screen. Viewing requires no special glasses or background/foreground screening.

Table 3.6 (*Continued*) Partial List of Volumetric Displays Available at Press Time

Jeff Han's Holodust	Involves infrared lasers "lighting up" particles in space, but it is not holographic. Holodust is essentially a swept-volume volumetric display, where the volume is a stochastic distribution of dustlike particles. The device works by rapidly sweeping an infrared laser throughout the display space. When the laser encounters a particle, some light is scattered and is picked up by high-speed position-sensing photodetectors. If that particle's position is deemed to be lit according to the model in voxel memory, a second laser, a visible one coaxial to the invisible one, is flashed on. When repeated rapidly with an appropriate distribution of particles, a coherent 3D image is formed.

3.4.1 The Accommodation Problem

With "normal view," two inverted images are formed on the retina at the back of the eye, with a parallax disparity. The brain takes in these two images and "fuses" them into a single image (a Cyclopean image, as discussed in Chapter 2) that appears to be seen from the center of the forehead. Objects in the Cyclopean image have depth proportional to the disparity between the left- and right-eye images. The brain actually "projects forward" the Cyclopean image, in the viewer's mind, to its correct position. When the viewer looks at objects, the viewer's two eyes turn inward (converge) to those objects. At the same time, the eyes focus (accommodate) on the point of convergence, in a control loop, to maximize the sharpness of the image on the retina. In simple planar 3D, the viewer is confronted with two images on a planar screen that have a disparity. The information the brain receives is that there are objects at different distances before it, and therefore it tries to get the eye to focus on them and to point to them. Unfortunately, to get the sharpest image in 3DTV, the eyes have to focus on the plane of the screen, and not where the brain thinks the objects are located in space. Focusing on the plane of the screen produces the sharpest images on the retina, but this is not where the objects "appear" to be; the brain is thus "confused," resulting in discomfort for the viewer if watched for long periods of time. Viewers can (eventually) train their brain to ignore some of the information and focus on the screen (that is, to separate the functions that are normally done together in concert) but the viewer is never completely comfortable, and this causes eye fatigue. (There are other causes of eye fatigue, some of which can be reduced by careful alignment and registration of the left and right pictures; this can be easier to do with digital systems, and thus such systems can have less eye fatigue.)

3.4.2 Infinity Separation

When objects are viewed at infinity, the two eyes have to point directly forward. To achieve this means that objects that are in the very far background (call that infinity distance) need to be displaced in the display by the same distance as the eyes are apart, that is about 65 mm. If the viewer has far-background images of a distance wider that 65 mm, this viewing becomes uncomfortable, because the two eyes have to point outwards; if the viewer has far-background images closer than 65 mm, the infinity moves forward, and the two eyes have to point inwards. It is possible to achieve proper calibration in a cinema where the two projectors or lenses can be adjusted so that the infinity distance is always at 65 mm; however, this is practically impossible to achieve in the home because the varying size of the TV displays in different instances makes it impossible to know before a transmission which difference in the transmitted signal will produce an absolute distance of 65 mm on a display.

3.5 Conclusion and Requirements of Future 3DTV

The following challenging observation may stimulate some thoughts on the matter of 3DTV service delivery in the short term [DOS200801]:

> Early demonstrations of stereoscopic color television or of stereoscopic film projection using polarization plane and color techniques to distinguish the content for the viewer's right eye from that of the left eye quickly revealed the problems faced when dealing with the human visual sense. A series of elements, termed "depth cues," contribute to our perception of depth. These include the relative size of the known-size objects, masking effects (when one object is in front of another), perspective—and most important of all is "binocular disparity." Binocular disparity is the difference between the same scene as seen by the left and right eyes, and is the most powerful depth cue for normally-sighted people. We might imagine that all we need to do to achieve 3D television is to arrange for "left-eye" and "right-eye" signals to be available

on a screen ("Planar 3D") separately to each eye of the viewer, and the job is almost done. True, but it may not be done "well." This applies whether the two pictures are delivered by analogue or by digital means. All planar 3D systems developed to date cause degrees of "eye fatigue." Some of the causes can be removed by such means as precise registration of the images for the left and right eye; but two causes in particular are not easily removed, whether the television system is analogue or digital. These are sometimes termed the "accommodation conflict" and the "infinity separation problem." They, and other problems, limit the possibilities for a long duration comfortable viewing 3D system for home television in the short term. Nevertheless, in spite of skepticism that a fatigue-free 3D television system can be made, it must be studied. It is a subject of great interest in broadcasting today. In addition, some movie companies are investing heavily in 3D as the potential savior of the "hall cinema" in the face of the growth of high-definition television (or HDTV) "home cinema." If fatigue-free 3D cinema can be found, then fatigue-free 3D television can also be found.

The 3D4YOU research done in Europe generated a set of requirements for a "quality" future 3DTV service. These requirements are included herewith to give the reader a sense of where this technology might be going in the next two to four years [KAU200801].

The principal goal of the 3D4YOU research done in Europe is to provide an autostereoscopic 3DTV service. A data representation format that uses "video-plus-depth" may be ideal for a future 3DTV system to provide such services. Figure 1.7 in Chapter 1 outlined a possible scenario of a future 3DTV system that is based on video-plus-depth concept. Different inputs can be used for 3D acquisition. This includes standard stereo cameras with two views, depth-range cameras directly providing additional depth information, and multi-baseline systems with more than two cameras. During 3D content production a suitable data representation format with video-plus-depth streams, MV+D in particular, is created from these inputs. The exact number N depends on the input material and the postproduction process. These streams can then be encoded either directly by a suitable MV+D coding profile, or can be used to create a new representation format, specifically, LDV. In LDV, the occlusion information is constructed by warping two or more neighboring video-plus-depth views from the MV+D representation onto a defined center view. The LDV stream or sub-streams can then be encoded by a suitable LDV coding profile. After decoding, the MV+D or LDV streams are converted to M regular video views by using depth-image-based rendering (DIBR). This conversion process depends on the properties and the number of views of the targeted 3D display and particular system configurations that adapts the 3D reproduction to local viewing conditions and individual user preferences. To this end, one can conclude that such a video-plus-depth approach can support the following features of an innovative 3DTV system design:

- Autostereoscopic multiuser displays are based on more than two stereo views. The exact number of display views depends on the particular 3D display type. Thus, the additional depth information can be used to convert a limited number of transmitted views to the needed number of display views by using depth-based rendering of intermediate views.
- Tracked autostereoscopic 3D displays allow continuous head-motion parallax viewing. Thus, the additional depth information can be used to render virtual stereo views continuously dependent on the current head position.
- During 3D production the interaxial camera distance is usually adjusted to a targeted 3D display platform with an assumed screen size and viewing distance. If these viewing conditions change, the additional depth information can be used for adapting the transmitted views to the new situation. This holds for both conventional stereo reproduction and autostereoscopic multiview displays. Thus, to achieve an optimal 3D impression, depth data can also be used to re-render an optimal virtual stereo view that depends on the local viewing conditions and the related user preferences.

In this context, it is one main objective of the 3D4YOU project to deliver an end-to-end system for 3D high-quality media. For this purpose it is necessary to define related requirements such as the types of 3D displays to be supported as well as further functionalities and viewing conditions:

- It can be supposed that all today's and future 3D displays that are relevant for 3DTV services are based on the principle of plano-stereoscopic devices; i.e., all available views will be displayed at the same planar screen. Thus, all relevant 3D displays will use horizontal parallax only. Other display technologies (integral imaging/photography, volumetric displays, holographic displays, etc.) that also provide vertical parallax or even volumetric reproductions are neither mature enough for short- or midterm solutions nor suitable for consumer applications due to too complex mechanics (scanning mirrors, rotating diffusers, etc.) and to insufficient spatiotemporal resolution.

- The rendering of intermediate views during 3D display adaptation should be as simple as possible. As much as possible preprocessing should already have been done in advance during capturing or postproduction. This particularly applies to the correction of lens distortions, normalization of intrinsic camera parameters, and rectification of convergent camera views. The processing at the receiver should be reduced to parallax scaling and shifting of pixels along horizontal scan lines according to the movement of a virtual camera along a common baseline.

- It is assumed that a user watches next-generation 3DTV from an almost static viewing position and that head and body only move in certain limits around this static position. If possible, 3D display adaptation should allow head motion parallax viewing in these limits; i.e., when watching 3DTV on a multiview display or a tracked stereo display, the user can slightly look behind objects by moving the head. However, this is limited to the usual range of horizontal head movement (approximately 50 cm).

- Interaction with the 3D content is limited to some basic features. The user can control depth reproduction by simple functions like scaling parallax and adding an offset. This is just for adapting depth sensation to user preferences and viewing conditions, comparable to control of brightness, contrast, and color at a regular TV set. Another more intuitive interaction is the provision of limited head motion parallax viewing as described above. More complex interactions like the active navigation with a virtual camera within the real 3D scene are not foreseen in the 3D4YOU scenario.

- Finally, the 3D representation and delivery format as well as 3D display adaptation should be backwards-compatible with existing services and systems. This applies especially to regular TV and DVB services. The implication here is that it should be possible to show one of the transmitted views directly at a conventional 2D display without transcoding or format conversion. In addition, it could be a consideration that the 3D representation is also compatible with stereo. However, the rationale here would not so much be legacy with existing transmissions (for stereo that is of a very different magnitude; most stereo transmissions are more for testing) but more for simplicity. Existing stereo systems, such as Samsung's 3D-Ready line, that are going to be established as stereo play-out systems for 3D home entertainment could be directly addressed without the need of DIBR. This, in particular, means that if two or more views are transmitted, they should preferably contain one pair of views that has been produced that way so that it can directly be displayed on a standard stereo display without any need of intermediate view interpolation or other conversion processes.

The main conclusions about a longer-term 3DTV system are therefore as follows:

- One main requirement of future 3DTV is that it has to support a wide range of different 3D displays, including conventional stereo with $M = 2$ views as well as autostereoscopic multiview display with $M > 2$ views. Today, the number of views at suitable 3D displays ranges from $M = 2$ to $M = 10$, but it can be expected that it will be much more in future. Thus, a given number N of transmitted views has to be converted to the particular number M of display views at the receiver side by means of image-based rendering. As a consequence, future 3DTV services will have to be based on video-plus-depth representation formats.
- Although 3D displays that are suitable for 3DTV applications will differ considerably with respect to the particular multiview reproduction technology, it can be assumed that they will exclusively be based on the geometry of plano-stereoscopic devices

where all views are presented at one common planar screen. Thus, to reduce the complexity of image-based rendering at the receiver side to its absolutely needed minimum, the transmitted views should strictly follow a parallel stereo geometry.

- The processing at the receiver required for 3D display adaptation should be as simple as possible. As much as possible processing should been done during capturing or postproduction processes. This particularly applies to the correction of colorimetric and geometric corrections. The processing at the receiver should be reduced to parallax scaling and shifting of pixels along horizontal scan lines accordingly to the movement of a virtual camera along a common baseline.

3DTV is viewed by some people as only a subset of an evolving 3D media. This is a characterization of those overarching goals [TEK200901]:

> With higher speed access network becoming widely available and affordable, services such as "Voice over IP," "video over IP," and "IPTV" are now commonly deployed and part of everyday life. 3D media services over IP are clearly the next big step forward in the evolution of digital media, entertainment, education, gaming and visual communication technologies. Significant accomplishments have been made over the last few years towards this goal; however, there are still some short-term and long-term fundamental research challenges that need to be addressed.

Figure 3.17, based largely on the same reference, depicts this evolving universe of 3D applications and technologies. New 3D video signal formats, capture technologies, and new-generation 3D displays will certainly emerge over the years. Multiview video may cause eyestrain and headache if not properly displayed; new 3D video capture and display technologies such as holoscopic and holographic video, discussed above, provide more natural 3D experience; however, these technologies are not mature enough for commercialization and more research is needed [TEK200901]. Compression of new video formats such as holographic video (dynamic holography) is of interest. New-generation networking technologies are also needed to support the service requirements of 3DTV, including bandwidth, latency, jitter, multigrade end-to-end quality of service (QoS) and multigrade application-level quality of experience (QoE). New 3D media Internet

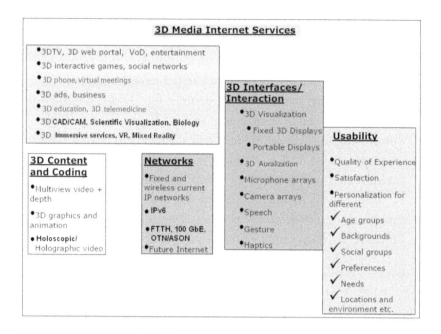

Figure 3.17   Universe of 3D media Internet services and enabling technologies.

services are also expected to emerge. Additionally, these services should become available globally, not only to Western countries. The O3B (Other 3 Billion) project aims at extending Internet, DTH, and other services to 3 billion people in the southern hemisphere not currently well served, using new innovative satellite and wireless services.*

References

[3DP200801] 3DPHONE, Project no. FP7–213349, Project title ALL 3D Imaging Phone, 7th Framework Programme, Specific Programme "Cooperation," FP7-ICT-2007.1.5—Networked Media, D5.2—Report on first study results for 3D video solutions, 31 December 2008.

[3DP200802] 3DPHONE, Project no. FP7–213349, Project title ALL 3D Imaging Phone, 7th Framework Programme, Specific Programme "Cooperation," FP7-ICT-2007.1.5—Networked Media, D5.1—Requirements and specifications for 3D video, 19 August 2008.

---

* For example, see http://www.o3bnetworks.com. The O3B Network's mission is to make the Internet accessible and affordable to everyone on the planet by reducing core transmission costs through innovative solutions for fixed and mobile operators and Internet service providers, enabling them to profitably offer better, faster, and more affordable connectivity to their customers.

[3DT200701] 3DTV, D26.2 3D time-varying scene capture technologies, 7 March 2007, *TC1 WP7 Technical Report 2*. C. Theobat, Ed. Project Number 511568, Project Acronym 3DTV.

[3DT200801] IST–6th Framework Programme, 3DTV NoE, 2004, L. Onural, project coordinator, IEEE Department, Bilkent University, TR-06800 Ankara, Turkey.

[3MO200901] MOBILE3DTV Project, 3D Media Cluster, Umbrella structure embracing related EC 3DTV funded projects, http://sp.cs.tut.fi/mobile3dtv

[AGG200601] A. Aggoun, Pre-processing of integral images for 3-D displays, *IEEE J. Display Technol.*, 2, 4, 393–400, 2006.

[AGG200901] A. Aggoun, 3D holoscopic imaging, 3D TV cameras, coding, monitors and application, Brunel University's Specialist Laboratories, *Pushing the Boundaries of 3D TV*, Newton Room, Brunel University, Uxbridge, Middlesex, UB8 3PH, 3 December 2009.

[ANA200901] A. Silliphant, http://www.anachrome.com/

[BAK200901] H. Baker, Z. Li, and C. Papadas, Calibrating camera and projector arrays for immersive 3D display, Hewlett-Packard Laboratories Papers, Palo Alto, CA, 2009, http://www.hpl.hp.com/personal/Harlyn_Baker/papers/BakerLiPapadas-EI-2009.pdf

[BCH200901] B.X. Chen, Wired explains: How 3D movie projection works, *Wired Online Magazine*, 21 December 2009.

[BOG198901] A. Bogusz, Holoscopy and holoscopic principles,, *J. Opt.* 20, 281–284, 1989.

[BOU200201] P. Bourke and D. Bannon, A portable rear projection stereoscopic display—A VPAC (Victorian Partnership for Advanced Computing) Project, Western Australian Supercomputer Program (WASP)/The University of Western Australia (UWA), April–October 2002.

[BRO200701] K. Brown, The use of integral imaging to realize 3D images, in true space, MIRIAD (Manchester Institute for Research & Innovation in Art & Design), Manchester Metropolitan University, UK, 2007.

[CHR200901] Chromatek Inc., brochure, Alpharetta, Georgia 30005.

[DAV199201] N. Davies and M. McCormick, "Holoscopic imaging with true 3D-content in full natural colour," J. Phot. Sci. 40, 46–49 (1992).

[DOS200801] C. Dosch and D. Wood, Can we create the "holodeck"? The challenge of 3D television, *ITU News Magazine*, no. 9, November 2008.

[FER200801] J. Fermoso, USC lab creates 3D holographic displays, brings TIE fighters to life, *Wired Magazine Online*, 26 June 2008.

[FUK200901] N. Fukushima, Calibration for image compensation required for 100-lens systems and also for handy Web stereo cameras, Report on the 3DC Conference 2009 in NAGOYA, Nagoya University, Venture Business Laboratory "Venture Hall" in Higashiyama Campus, 12 June 2009.

[HDH200701] HDHTV, High definition holographic television, HDHTV c/o Virtual Search, Minnetonka, MN, 2007.

[HOL201001] Holografika, Real 3D display by Holografika, promotional material.

[JUN200201] S. Jung, J-H. Park, H. Choi, and B. Lee, Implementation of wide-view integral 3D imaging using polarization switching, Lasers and Electro-Optics Society, 2002 (LEOS 2002), *15th Annual Meeting of the IEEE 2002*, 1, 31–32.

[KAU200801] P. Kauff, M. Müller, et al., ICT-215075 3D4YOU, Deliverable D2.1.2: Requirements on post-production and formats conversion, August 2008.

[MAR200601] M. Martínez-Corral, R. Martínez-Cuenca, G. Saavedra, and B. Javidi, Integral imaging: Autostereoscopic images of 3D scenes, *SPIE Newsroom*, 2006.

[MAR200801] M. Martínez-Corral, R. Martínez-Cuenca, G. Saavedra, H. Navarro, A. Pons, and B. Javidi, Progresses in 3D integral imaging with optical processing, *J. Phys.: Conf. Ser.* 139, 2008.

[MAS200901] K. Mase, Research on usage of multi-view video for production skill succession, Report on the 3DC Conference 2009 in NAGOYA, Nagoya University, Venture Business Laboratory "Venture Hall" on Higashiyama Campus, 12 June 2009.

[MER200901] R. Merritt, Incomplete 3DTV products in CES spotlight HDMI upgrade one of latest pieces in stereo 3D puzzle, *EE Times*, 23 December 2009.

[MIN200501] D. Minoli, *Nanotechnology Applications to Telecommunications and Networking*, Wiley, Hoboken, NJ, 2005.

[MON200101] S. Monaleche, A. Aggoun, A. McCormick, N. Davies, and Y. Kung, Analytical model of a 3D recording camera system using circular and hexagonal based spherical microlenses, *J. Optical Soc. America A*, 18, 8, 1814–1821, 2001.

[ONU200601] L. Onural, A. Smolic, et al., An assessment of 3DTV technologies, *2006 NAB BEC Proceedings*, 456ff.

[ONU200801] L. Onural and H. M. Ozaktas, Three-dimensional television: From science-fiction to reality, in *Three-Dimensional Television: Capture, Transmission, Display*, H. M. Ozaktas.; L. Onural, Eds., Springer Verlag, New York, 2008.

[PAT200901] M. Patkar, How 3DTV works: Part I—With glasses, *Thinkdidit. com Online Magazine*, 20 October 2009.

[PAT200902] M. Patkar, How 3DTV works: Part II—Without glasses, *Thinkdidit.com Online Magazine*, 26 October 2009.

[POW200301] E. Powell, What is an anamorphic lens? Projector Central, Portola, CA, 18 September 2003.

[REA200901] Real3D Project, 3D media cluster, Umbrella structure embracing related EC 3DTV funded projects, http://www.digitalholography.eu/

[STA200901] M. Starks, SPACESPEX™ anaglyph—The only way to bring 3DTV to the masses, Online article, 2009.

[STA200801] M. Starks, *Digital 3D Projection*, 3DTV Corp., Springfield, OR, 2008.

[SUN200901] D. Sung, 3D dictionary, *Gadget News, Reviews, Videos*, http://www.pocket-lint.com, 21 September 2009.

[TAK200201] Y. Takaki, Three-dimensional display with 64 horizontal parallaxes, Dept. of Electrical and Electronic Engineering Tokyo University of
Agriculture and Technology, Koganei, Tokyo, Japan, August 2002.

[TEC200901] Technest/Genex Technologies, Inc., Displays for 3D imaging,
3D volume viewer—The glasses-free walk-around display, Bethesda, MD.

[TEK200901] A.M. Tekalp, 3D media delivery over IP, *IEEE Multimedia
Commn. Tech. Comm. E-Letter*, 4, 3, April 2009.

[TSE200901] E. Tsekleves, J. Cosmas, A. Aggoun, and J. Loo, Converged digital TV services: The role of middleware and future directions of interactive television, *Int. J. Digital Multimedia Broadcasting*, 2009.

[ZYG200801] L. Zyga, Improved volumetric displays may lead to 3D computer monitors, PhysOrg.com, 22 December 2008.

4

Basic 3DTV Approaches for Content Capture and Mastering

This chapter looks at the basic subsystems that comprise an overall 3DTV system, including subsystems for capture, mastering, and distribution. We assume here that the commercial 3DTV under discussion is the conventional stereo video (CSV) model and not the more advanced, but further in the future, multiview/free-viewpoint/autostereoscopic system described at the end of Chapter 3; however, many of the elements discussed here would also apply to the more advanced models. Appendix 4A provides some additional insights on 3D video formats.

4.1 General Capture, Mastering, and Distribution Process

Figure 4.1 depicts a general end-to-end view of a 3DTV signal management system. The key functional components of 3DTV are capture and representation of 3D scene information, complete definition of digital 3DTV signal, storage and transmission of this signal, and finally, display of the reproduced 3D scene. For a successful consumer-accepted operation of 3DTV, all these functional components must be designed in an integrated fashion; this kind of large-scale integration is highly multidisciplinary and thus requires cross-industry cooperation [3DT200801]. From an implementation perspective, the basic equipment supporting the overall process flow (capture, mastering, and distribution) is shown in basic form in Figure 4.2.

Content can be captured directly in digital form or on film. Movies, whether originally shot digitally or on traditional film, are transferred to a digital intermediate format for any postproduction editing or production-level storage. During the postproduction phase of a film-based movie, it is common to scan the camera original negative into a computer

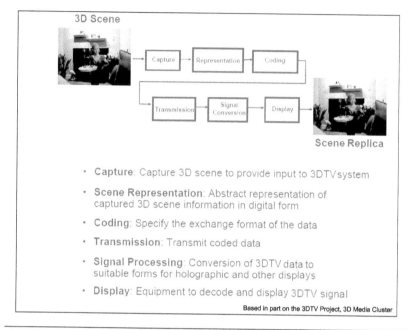

Figure 4.1 (Please see color insert following page 160) End-to-end signal management—general view.

file system where each picture becomes a digital data file. This conversion can be carried out at a variety of resolutions, typically at 4K—that is, 4,096 picture elements (pixels) horizontally × 3,112 picture elements vertically; this can also be done at 2K, or 2,048 elements horizontally × 1,556 elements vertically. Typical file formats at this postproduction stage are Kodak Cineon or digital picture exchange (DPX).

Mastering (also known as *content preparation*) is the process of creating the master file package containing the movie's images, audio, subtitles, and metadata. Mastering standards are typically used in this process. Generally, mastering is performed on behalf of the distributor/broadcaster at a mastering facility. Encryption may be added at this stage. At this juncture the file package is ready for duplication or for transmission via satellite or fiber links. The content is then either printed to regular film-strip reels for playback in conventional motion-picture theaters or sent out on hard drives to theaters that project movies digitally. From the 2K–5K digital masters, the studios can then make 480 standard-definition DVDs or 1,080 high-definition discs. Older films are being archived at 2K–5K as well, and then down-converted to DVD or Blu-ray.

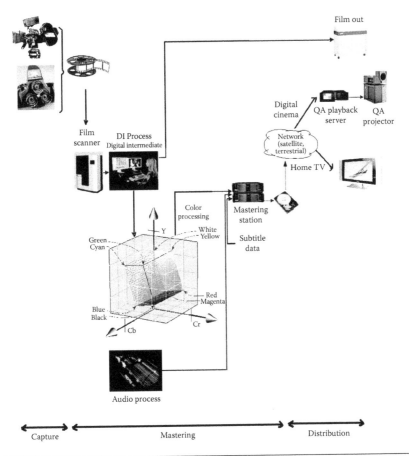

Figure 4.2 (Please see color insert) Overall video capture/mastering/distribution process—implementation perspective.

Some examples of mastering follow for illustrative purposes:

- Digital cinema mastering is the process of converting the digital intermediate (DI) film-out (image) files into compressed, encrypted track files, and then combining these image track files with the uncompressed audio track files and subtitle track files to form a DCI/SMPTE-compliant digital cinema. Hence the term describes a system of data reduction, then reel building with audio and subtitles, security, and packaging. After the scanning, the data enters into the DI process, which can add computer graphics and animations, special effects, and color correction. The DI will be tuned to output to film recorders and a second output will go through a process commonly called *color cube* * *for digital cinema* and possibly other digital releases. A color cube is also put in the path to the grading projector to simulate the film. A color cube is a multidimensional lookup table used to modify the red, green, and blue values; the numbers in the cube are normally proprietary. This data file is called a digital source master (DSM). The DSM may be further preprocessed to obtain the digital cinema distribution master (DCDM); the reformatted file is the DCDM and is normally in the form of a tagged image file format (TIFF) file. For example, projector resolutions are only 8.8 megapixels for 4K and 2.2 megapixels for 2K, and so the DSM needs to be appropriately customized for the intended application at hand [EDC200701].

• The DVD mastering process is the process whereby a film source with an aspect ratio greater than 4:3 is transferred to the DVD video master in such a way that the picture is vertically stretched by a factor of about 1.33 (e.g., if the picture had an aspect ratio of 16:9, it now has one of 4:3). The goal is to retain as much resolution of the video master as possible so that widescreen pictures use the 4:3 frame optimally, gaining another 33 percent of vertical resolution and look sharper. When playing a DVD with an anamorphic widescreen, the display (16:9 capable TV or projector and screen) has to vertically squeeze the picture by a factor of 0.75 (so that, for example, a circle is still a circle, or a square still a square). If the display cannot do this, the DVD player will do the squeezing and add black bars on the top and bottom of the picture. In that case, the additional 33 percent resolution is not available [IMD201001].

• 3D mastering, as being defined by the Society of Motion Picture and Television Engineers (SMPTE) (discussed below in Section 4.2.2).

---

* A color cube is a three-dimensional representation of color. Its function in the DI suite is to modify the film information sent to the film recorder to match the film characteristics to the colorist's view. It is basically a 3D lookup table in hardware or software. The numbers are a prized secret between postproduction houses.

4.2 3D Capture, Mastering, and Distribution Process

An end-to-end 3DTV system is illustrated in Figure 4.3. The capture, mastering, and distribution approaches for 3D are an extension of traditional video capture approaches. We discuss the substeps next. Some have argued that in future 3DTV systems' scene capture and display operations may be decoupled: captured scene information may be converted to abstract representations (and maybe stored), potentially using computer graphics techniques, and the display (and observer) will interact with this intermediate data [ONU200601]. However, in the short term more basic CSV methods will be employed.

4.2.1 Content Acquisition

Real-time capture of 3D content almost invariably requires a pair of cameras to be placed side-by-side in what is called a 3D rig to yield a left-eye/right-eye view of a scene* (see Figure 1.5). The lenses on the left and right cameras in a 3D rig must match each other precisely. Furthermore, the precision of the alignment of those two cameras is critical; misaligned 3D video is cumbersome to watch and will be stressful to the eyes.

Concepts such as focal length, iris, focus, interaxial separation, and convergence are key concepts, among others (see Appendix 2A for further discussion of some of these elements).

The focal point of a lens is the point where the image of a parallel, entering bundle of light rays is formed. The focal length (usually designated by f') is the distance from the surface of a lens or mirror to its focal point (that is, the distance of the focal point from the lens); it is a measure of the collecting or diverging power of a lens or an optical system. The iris is a diaphragm consisting of thin overlapping plates that can be adjusted to change the diameter of a central opening; it is a mechanical device in a camera that controls size of aperture of the lens.

Two parameters of interest for 3D camera acquisition are *camera separation* and *toe-in*. This operation is similar to how the human eyes work: as one focuses on an object in close proximity, the eyes toe-in;

---

* Content acquired in 3D can be delivered in 2D by using the images from only one of the cameras, usually the left-eyed one; the 3D-ready material can be published at a future time.

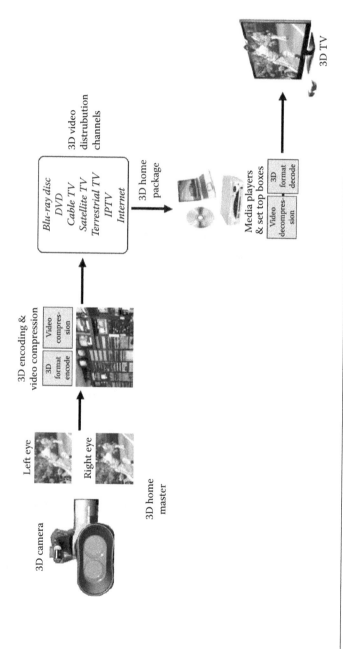

Figure 4.3   End-to-end 3DTV system.

as one focuses on remote objects, the eyes are parallel (as discussed in Chapter 2). Interaxial distance t_e (also known as interaxial separation) is the distance between camera lenses' axes; this can be also defined as the distance between two positions for a stereo photograph. The baseline distance between visual axes (separation) for the eyes is around 2.5 inches (65 mm). 3D cameras use the same separation for baseline, but the separation can be smaller or larger to accentuate the 3D effect of the displayed material. The term interocular (IO) is also used for the distance between the two cameras (or lenses). As noted, that distance determines how extreme the 3D effect is and also will need to be varied for different focal length lenses and by the distance from cameras to the subject (recall the discussion on Z_{near} and Z_{far} in Appendix 2A). 3D rigs provide for interocular adjustment. The two cameras in the 3D rig are angled slightly toward each other so that their framing matches up exactly at some distance away. That adjustment is known as convergence. In 3D, objects in the frame at that convergence point will appear to be located at the screen. Objects closer to the camera than that convergence point will appear to be located in front of the screen, and those farther away will appear to be behind the screen [JOH200901]. With video capture, one can check alignment, IO, and convergence in real time and correct these parameters before one shoots the scene; for multicamera shoots, IO and convergence can be matched among the various camera pairs and can be adjusted in real time during the shoot. When shooting 3D on film, one typically aligns the two cameras mechanically, but it is not until one gets the processed film back for inspection that one can be sure that the alignment is correct.

We have mentioned all along that the availability of content will be critical to the successful introduction of the 3DTV service, and that 3D content is more demanding in terms of production. Material created for cinema screens may have to be scaled down for TV screens and the production of real-time events is currently a challenge, although doable. Some content owners are considering the creation of 3D material by converting a 2D movie to a stereoscopic product with left/right eye tracks; in some instances, non-real-time conversion of 2D to 3D may lead to marginally satisfactory results. The fact remains, however, that it is not straightforward to create a stereopair from 2D content (issues relate to object depth and reconstruction of parts of the image that are obscured in the first eye). Nonetheless, conversion from 2D may play a role in the short term.

A number of measures should be taken to reduce eye fatigue in 3D during content development and creation. These include restricting depth of scene, positioning the key objects in the plane of the screen, and having a restricted lens separation to near-object distance ratio. The entire 3D scene needs to be "in focus" from front to back. In addition, content developers are challenged to find ways to create "real-life" 3D. A simple planar 3D system always generates the same depth cues regardless of the position of the viewer's head (or body), but this is not the case in real life where disparity changes with viewer's head position [DOS200801]. To fully address all issues one would need to rely on holographic solutions (as covered in Chapter 3), but these are not anticipated to be available in the near future.

4.2.2 3D Mastering

A 3D mastering standard called "3D master" is being defined by SMPTE. The high-resolution 3D master file is that used to generate other files appropriate for various channels, such as theater release, media (DVD, Blu-ray disc) release, and broadcast (satellite, terrestrial broadcast, cable TV, IPTV, and/or Internet distribution). The 3D master is comprised of two uncompressed files (left and right eye), each of which has the same file size as a 2D video stream. Formatting and encoding procedures have been developed that can be used in conjunction with already-established techniques to deliver 3D programming to the home over a number of distribution channels.

In addition to normal video encoding, 3D mastering/transmission requires additional encoding/compression, particularly when attempting to use legacy delivery channels. Encoding schemes include the following:

- Spatial compression
- Temporal multiplexing
- 2D in conjunction with metadata
- Color encoding

4.2.2.1 Spatial Compression When an operator seeks to deliver 3D content over a standard video distribution infrastructure, spatial

Figure 4.4 (Please see color insert) Spatial compression.

compression is a common solution. Spatial compression allows the operator to deliver a stereo 3D signal (now called frame compatible) over a 2D high-definition (HD) video signal, making use of the same amount of channel bandwidth. Clearly this entails a loss of resolution (for both the left and the right eyes). The approach is to pack two images into a single frame of video; the receiving device (for example, the set-top box) will in turn display the content in such a manner that a 3D effect is perceived (these images cannot be viewed on a standard 2D TV monitor). There are a number of ways of combining two frames; the two most common are the side-by-side combination and the over/under combination.* See Figure 4.4, where the two images are reformatted at the compression/mastering point to fit into that standard frame. The combined frame is then compressed by standard methods and delivered to 3D-compatible TV, where it is reformatted/rendered for 3D viewing.

The question is how to take two frames, a left frame and a right frame, and reformat them to fit side by side or over/under in a single standard HD frame. Sampling is involved but, as noted, with some loss of resolution (50 percent). One approach is to take alternate columns of pixels from each image and then pack the remaining columns

---

* This is also called "top/bottom" by some.

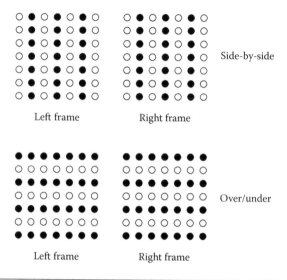

Side-by-side

Left frame   Right frame

Over/under

Left frame   Right frame

Figure 4.5 Selection of pixels in side-by-side and over/under approaches.

in the side-by-side format. Another approach is to take alternate rows of pixels from each image and then pack the remaining rows in the over/under format (see Figure 4.5).

Studies have shown that the eye is less sensitive to loss of resolution in a diagonal direction of an image than it is in the horizontal or vertical direction. This allows the development of encoders that optimize subjective quality by sampling each image in a diagonal direction. Other encoding schemes are also being developed to attempt to retain as much of the perceived/real resolution as possible.

One approach that has been studied for 3D is quincunx filtering. A quincunx is a geometric pattern comprised of five coplanar points, four of them forming a square (or rectangle) and a fifth point at its center, like a checkerboard. Quincunx filter banks are 2D two-channel nonseparable filter banks that have been shown to be an effective tool for image-coding applications. In such applications, it is desirable for the filter banks to have perfect reconstruction, linear phase, high coding gain, good frequency selectivity, and certain vanishing-moment properties (for examples, see [TAY199301], [SWE199601], [GOU200001], [KOV200001], [CHE200701], [LIU200601] among others). Almost all hardware devices for digital image acquisition and output use square pixel grids. For this reason and for the ease of computations, all current image-compression algorithms (with the

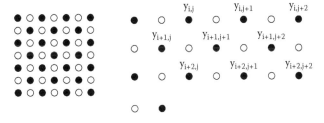

Figure 4.6 Quincunx sample grid. Note: Either black or white dots can comprise the lattice.

exception of mosaic image compression for single-sensor cameras) operate on square pixel grids. The optimal sampling scheme in the two-dimensional image space is claimed to be the hexagonal lattice; unfortunately, a hexagonal lattice is not straightforward in terms of hardware and software implementations. A compromise therefore is to use the quincunx lattice; this is a sublattice of the square lattice, as illustrated in Figure 4.6. The quincunx lattice has a diamond tessellation that is closer to optimal hexagon tessellation than the square lattice, and it can be easily generated by down-sampling conventional digital images without any hardware change. Because of this, quincunx lattice is widely adopted by single-sensor digital cameras to sample the green channel; also, quincunx partition of an image was recently studied as a means of multiple-description coding [ZHA200701].

When using quincunx filtering, the higher-quality sampled images are encoded and packaged in a standard video frame (either with the side-by-side or over/under arrangement). The encoded and reformatted images are compressed and distributed to the home using traditional means (cable, satellite, terrestrial broadcast, and so on).

4.2.2.2 Temporal Multiplexing Temporal (time) multiplexing doubles the frame rate to 120 Hz to allow the sequential repetitive presentation of the left-eye and right-eye images in the normal 60 Hz time frame. This approach retains full resolution for each eye but requires a doubling of the bandwidth and storage capacity.

In some cases spatial compression is combined with time multiplexing; however, this is more typical of an in-home format and not a transmit/broadcast format. As an example, Mitsubishi's 3D DLP TV uses quincunx sampled (spatially compressed) images that are clocked at 120 Hz as input.

4.2.2.3 2D in Conjunction with Metadata (2D+M) The basic concept here is to transmit 2D images and to capture the stereoscopic data from the "other eye" image in the form of an additional package, the metadata; the metadata is transmitted as part of the video stream. For example, metadata can be a depth map (see Figure 4.7). This approach is consistent with Moving Picture Experts Group (MPEG) multiplexing, and therefore to a degree it is compatible with embedded systems. The requirement to transmit the metadata increases the bandwidth needed in the channel: the added bandwidth ranges from 60 to 80 percent depending on quality goals and techniques used. As implied, a set-top box employed in a traditional 2D environment would be able to use the 2D content, ignoring the metadata, and properly display the 2D image; in a 3D environment, the set-top box would be able to render the 3D signal.

Some variations of this scheme have already appeared. One approach is to capture a delta file that represents the difference between the left and right images. A delta file is usually smaller than the raw file because of intrinsic redundancies. The delta file is then transmitted as metadata. Companies such as Panasonic and TDVision use this approach, which can also be used for stored media. For example, Panasonic has advanced (and the Blu-ray Disc Association is studying) the use of metadata to achieve a full-resolution 3D Blu-ray disc standard. A 1920x1080p 24 fps resolution per eye is achievable. This standard would make Blu-ray disc a high-quality 3D content (storage) system. The goal was to agree to the standard by early 2010 and have 3D Blu-ray disk players emerge by the end-of-year shopping season 2010.

Another approach (advanced by Philips) entails transmitting the 2D image in conjunction with a depth map of each scene, as noted above.

As noted elsewhere, many 3DTV proposals often rely on the basic concept of "stereoscopic" video, that is, the capturing, transmission, and display of two separate video streams (one for the left eye and one for the right eye). More recently specific proposals have been made for a flexible joint transmission of monoscopic color video and associated per-pixel depth information (see, for example, [FEH200201], [FEH200301]). From this data representation, one or more "virtual" views of the 3D scene are then generated in real time at the receiver side by means of depth-image-based rendering (DIBR) techniques, as discussed in Section 3.5 [FEH200601]. A system such as this

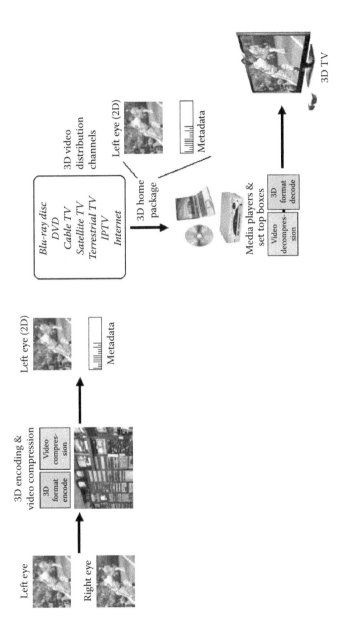

Figure 4.7 2D in conjunction with metadata.

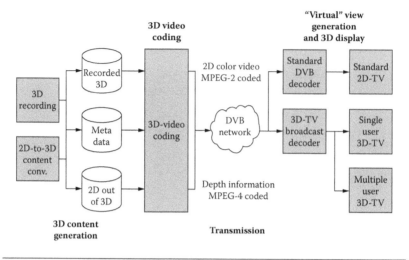

Figure 4.8   Depth-image-based rendering (DIBR) system.

provides important features, including backwards-compatibility to today's 2D digital TV, scalability in terms of receiver complexity, and easy adaptability to a wide range of different 2D and 3D displays. DIBR is the process of synthesizing "virtual" views of a scene from still or moving color images and associated per-pixel depth information. Conceptually, this novel view generation can be understood as the following two-step process: At first, the original image points are reprojected into the 3D world, utilizing the respective depth data; thereafter, these 3D space points are projected into the image plane of a "virtual" camera that is located at the required viewing position. The concatenation of reprojection (2D to 3D) and subsequent projection (3D to 2D) is usually called 3D image warping in the computer graphics (CG) literature and will be derived mathematically in the following paragraph. The signal processing and data transmission chain of this kind of 3DTV concept is illustrated in Figure 4.8; it consists of four different functional building blocks: (1) 3D content creation, (2) 3D video coding, (3) transmission, and (4) "virtual" view generation and 3D display.

In summary, broadcasters appear to be rallying around the top/bottom (over/under) approach; however, trials were still ongoing at press time. Other approaches involving some form of compression include checkerboard (quincunx filtering), side-by-side, or interleaved rows or columns [MER200902].

4.2.2.4 Color Encoding Color encoding (anaglyph) is the *de facto* method that has been used for 3D over the years. There are many hundreds of patents on anaglyphs in a dozen languages going back 150 years. The left-eye and right-eye images are color encoded to derive a single merged (overlapped) frame; at the receiving end, the two frames are restored (separated) using colored glasses. This approach makes use of a number of encoding processing techniques to optimize the signal in order to secure better color contrast, image depth, and overall performance (see Figure 4.9 and Figure 4.10). Red/blue, red/cyan, green/magenta, and blue/yellow color coding can be used, with the first two being the most common. Orange/blue anaglyph techniques are claimed by some to provide good quality, but there is a continuum of combinations [STA200901]. Advantages of this approach are that it is frame compatible with existing systems, can be delivered over any 2D system, provides full resolution, and uses inexpensive "glasses." However, it produces the lowest quality 3D image compared with the other systems discussed above. This approach is *not* likely to be implemented at this time for the commercial deployment of 3DTV.

Appendix 4A provides some additional insights on 3D video formats.

4.3 Overview of Network Transport Approaches

In many countries it is expected that 3DTV for home use will likely first see penetration via stored media delivery (such as Blu-ray disc). However, in several Western countries one saw commercial real-time 3DTV transmission services starting in 2010 and continuing to expand from that point forward. The broadcast commercial delivery of 3DTV (whether over satellite/DTH, over the air, over cable, or via IPTV) may take some number of years because of the relatively large-scale infrastructure that has to be put in place by the service providers and the limited availability of 3D-ready TV sets in the home (implying a small subscriber and therefore small revenue base). Delivery of downloadable 3DTV files over the Internet may occur at any point in the immediate future, but the provision of a broadcast-quality service over the Internet is not likely for the foreseeable future.

There are generally two potential approaches for transport of 3DTV signals: (1) connection-oriented (time/frequency division multiplexing)

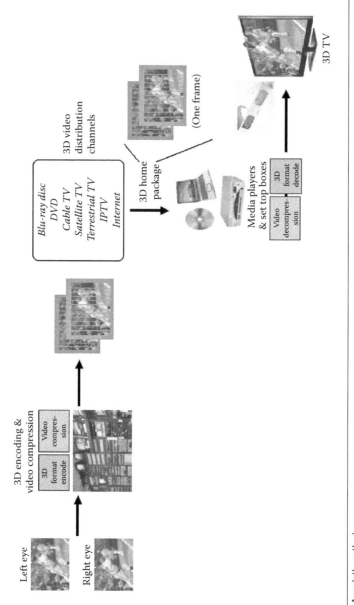

Figure 4.9 Anaglyth method.

Figure 4.10 (Please see color insert) Anaglyth method—another view.

over existing DVB infrastructure over traditional channels (such as sat-ellite, cable, over-the-air broadcast, digital video broadcasting-handheld [DVB-H]/cellular), and (2) connectionless/packet using the Internet protocol (IP) (such as "private/dedicated" IPTV networks, Internet streaming, Internet on-demand servers/P2P). These references, among others, describe various methods for traditional video over packet/ATM/IPTV/satellite/Internet [MIN200901], [MIN200801], [MIN199501] [MIN199401] [MIN200001] [MIN199601]; many of these approaches and techniques can be extended, adapted, or used for 3DTV.

The take-away here is that there are a number of alternative trans-port architectures for 3DTV signals, also depending on the underlying media. The service can be supported by traditional broadcast structures, including the DVB architecture, wireless 3G/4G transmission such as DVB-H approaches, IP in support of an IPTV-based service (in which case, it also makes sense to consider IPv6), and Internet-based delivery (both non-real time and streaming). The specific approach used by each of these transport methods will also depend on the video capture approach, as depicted in Table 4.1. Also, in the United States there is a well-developed cable infrastructure in all Tier 1 and Tier 2 metropolitan and suburban areas; in Europe and Asia, this is less prevalent, with more DTH delivery (in the United States, DTH tends

Table 4.1   Video Capture and Transmission Possibilities

	TERRESTRIAL DVB	DTH WITH DVB	3G/4G + DVB-H	IPTV (IPV4 OR IPV6)	INTERNET REAL-TIME STREAMING	INTERNET NON-REAL-TIME	CABLE TV
Conventional stereo video (CSV)	fine	fine	limited	fine	fine	fine	fine
Video plus depth (V+D)	good	good	fine	good	good	good	good
Multiview video plus depth (MV+D)	good	good	fine	good	fine	good	good
Layered depth video (LDV)	best	best	fine	best	fine	good	best

Note: Fine = doable; good = better approach; best = best approach to serve more exurban and rural areas). A 3DTV rollout must take these differences into account and/or accommodate both.

Note that the V+D data representation can be utilized to build 3DTV transport evolutionarily on the existing DVB infrastructure. The in-home 3D images are reconstructed at the receiver side by using DIBR; to that end, MPEG has established a standardization activity that focuses on 3DTV using video-plus-depth representation.

A challenge in the deployment of multiview video services, including 3D and free-viewpoint TV, is the relatively large bandwidth requirement associated with transport of multiple video streams. Two-stream signals [conventional stereo video (CSV), video plus depth (V+D), and layered depth video (LDV)] are doable—the reliable delivery of a single stream of 3D video in the range of 20 Mbps is not outside the technical realm of most providers these days—but to deliver a large number of channels in an unswitched mode (requiring say 2 Gbps access to a domicile) will require fiber to the home (FTTH) capabilities. It is not possible to deliver that content over an existing copper plant of the xDSL nature unless a provider could deploy ADSL2+. (But why bother upgrading a plant to a new copper technology such as this one when the provider could actually deploy fiber?) However, ADSL2+ may be used in multiple dwelling units as a riser for a FTTH plant. A way to deal with this is to provide user-selected multicast capabilities where a user can select an appropriate content channel using IGMP (Internet group management protocol). Even then, however, a household may have multiple TVs (say, three or four) going simultaneously (and maybe even an active DVR), thus requiring bandwidth in the 15–60 Mbps range. Multiview video plus depth (MV+D), where one wants to carry three or even more intrinsic (raw) views, becomes much more challenging and problematic for practical commercial applications.

Off-the-air broadcast could be accomplished with some compromise by using the entire HDTV bandwidth for a single 3DTV channel—here multiple TVs in a household could be tuned to different such programs.

A traditional cable TV plant would likely be challenged to deliver a large pack of full-resolution 3DTV channels—but they could deliver a subset of their total selection in 3DTV (say, 10 or 20 channels) by sacrificing bandwidth on the cable that could otherwise carry distinct channels or by using spatial compression. The same is true for DTH applications.

For IP, a service-provider-engineered network could be used. Here the provider can control the latency, jitter, effective source–sink bandwidth, packet loss, and other service parameters. However, if the approach is to use the Internet, performance issues will be a major consideration, at least for real-time services. A number of multiview encoding and streaming strategies using real-time transport protocol/user datagram protocol/Internet protocol (RTP/UDP/IP) or real-time transport protocol/datagram congestion control protocol/Internet protocol (RTP/DCCP/IP) exist for this approach. Video-streaming architectures can be classified as (1) server to single client unicast, (2) server multicasting to several clients, (3) peer-to-peer (P2P) unicast distribution, where each peer forwards packets to another peer, and (4) P2P multicasting, where each peer forwards packets to several other peers. Multicasting protocols can be supported at the network layer or application layer [D32200701].

While it is likely that initially 3DTV will be delivered by traditional transport mechanisms, including DVB over DTH systems, recently some research efforts have been focused on delivery (streaming) of 3DTV using IP. IP can be used for IPTV systems or over an IP shared infrastructure, whether a private network (here shared with other applications) or over the Internet (here shared with a multitude of other users and applications). Some studies have also been undertaken of late on the capabilities of DVB-H to broadcast stereovideo streams. However, it seems that the focus so far has been on IP Version 4 (IPv4); the industry is encouraged to assess the capabilities of IP Version 6 (IPv6) [MIN200802]. Although this topic is partially tangential to a core 3DTV discussion, the abundant literature on proposals for packet-based delivery of future 3DTV makes the issue relevant. IPv6, when used with header compression, is expected to be a very useful technology to support IPTV in the future in general and 3DTV in particular.

In summary, real-time delivery of 3DTV content can make use of satellite, cable, broadcast, IP, IPTV, Internet, and wireless technologies. Any unique requirements of 3DTV need to be taken into account; generally speaking the requirements are very similar to those needed for delivery of entertainment-quality video (such as in reference to latency, jitter, packet loss, and so on) but with the exception that a number (if not most) of the encoding techniques require more bandwidth. The incremental bandwidth is as follows: (1) from 20 to 100 percent more for stereoscopic viewing compared with 2D viewing;* (2) from 50 to 200 percent more for multiview systems compared with 2D viewing; and (3) a lot more bandwidth for holoscopic/ holographic designs (these not even presently being considered for near-term commercial 3DTV service).

4.4 MPEG Standardization Efforts

Standards are key to the cost-effective deployment of a technology. Examples of video-related standards include the Beta–VHS and the HD DVD–Blu-Ray controversies.† In order to roll out 3DTV services on a broad scale, standards have to be available. Standards need to cover many, if not all, elements depicted in Figure 4.1, including capture, mastering, distribution, and consumer device interface. Standards for 3D transport issues are particularly important because content providers and studios seek to create one master file that can carry stereo 3D content (and 2D content by default) across all the various distribution channels, including cable TV, satellite, over the air, packaged media, and the Internet. It will probably be somewhere around 2012 by the time that there will be an interoperable standard available in consumer systems to handle all the delivery mechanisms for 3DTV.

As noted in Chapter 1, there is a lot of industry and standards body interest in this topic. The MPEG of ISO/IEC is working on a coding format for 3D video. As we alluded to in Chapter 1, SMPTE is working on some of the key standards needed to deliver 3D to the home. At a broad level and in the context of 3DTV, the following major initiatives were afoot at press time:

---

* This is for V+D approaches; reduced-resolution methods using spatial compression do not require extra bandwidth.
† HD DVD (High-Definition/Density DVD) was a high-density optical disc format for storing data and high-definition video advanced principally by Toshiba. In 2008 after a protracted format war with rival Blu-ray, the proposed format was abandoned.

- MPEG, standardizing multiview and 3D video coding
- DVB, standardizing of digital video transmission to TVs and mobile devices
- SMPTE, standardizing 3D delivery to the home
- ITU-T, standardizing user experience of multimedia content
- Video Quality Experts Group (VQEG), standardizing of objective video quality assessment

We review here some of the ongoing MPEG standardization work in this section.

Simulcast coding is the separate encoding (and transmission) of the two video scenes in the conventional stereo video (CSV) format; clearly the bit rate will typically be in range of double that of 2DTV. Video plus depth (V+D) (also called 2D plus depth, or 2D+depth, or color plus depth) is more bandwidth efficient not only in abstract but also in practicality. At the pragmatic level, in a V+D environment, the quality of the compressed depth map is not a significant factor in the final quality of the rendered stereoscopic 3D video; this follows from the fact that the depth map is not directly viewed but is employed to warp the 2D color image to two stereoscopic views. Studies show that the depth map can typically be compressed to 10–20 percent of the color information.

The H.264/AVC (advanced video coding) standard was enhanced a few years back with a stereo supplemental enhancement information (SEI) message that can also be used to implement a prediction capability that reduces the overall bandwidth requirement. Although not designed for stereo-view video coding, the H.264 coding tools can be arranged to take advantage of the correlations between the pair of views of a stereo-view video and provide very reliable and efficient compression performance as well as stereo/monoview scalability.

The concept of a scalable video-coding scheme is to enable the encoding of a video stream that contains one (or several) subset bit stream(s) of a lower spatial or temporal resolution (that is, lower quality video signal)—each separately or in combination—compared to the bit stream it is derived from (for example, the subset bit stream is typically derived by dropping packets from the larger bit stream) that can itself (themselves) be decoded with a complexity and reconstruction quality comparable to that achieved using the existing coders (such as H.264/MPEG-4 AVC) with the same quantity of data as in the subset bit stream. A scalable video coding as specified in Annex G of H.264/AVC was added in 2007. With the help of the stereo video SEI message defined in H.264 fidelity range extensions (FRExt), a decoder can easily synchronize the views, and a streaming server or a decoder can easily detect the scalability of a coded stereo video bit stream [SUN200501].

Video plus depth (V+D) has been standardized in MPEG as an extension for 3D filed under ISO/IEC FDIS 23002-3:2007(E). In 2007 MPEG specified a container format "ISO/IEC 23002-3 Representation of Auxiliary Video and Supplemental Information" (also known as MPEG-C Part 3) that can be utilized for V+D data. 2D+Depth, as specified by ISO/IEC 23002-3, supports the inclusion of depth for generation of an increased number of views. While it has the advantage of being backward compatible with legacy devices and is agnostic of coding formats, it is only capable of rendering a limited depth range since it does not directly handle occlusions [ISO200901]. Transport of this data is defined in a separate MPEG systems specification, "ISO/IEC 13818-1:2003 Carriage of Auxiliary Data."

There is also major interest in MV+D. It has been recognized that multiview video coding is a key technology for a wide variety of future applications, including FVV/FTV (free-viewpoint video/free-viewpoint television), 3DTV, immersive teleconference and surveillance, and other applications. A MPEG standard, multiview video coding (MVC), to support MV+D (and also V+D) encoded representation inside the MPEG-2 transport stream has been developed by the Joint Video Team (JVT) of ISO/IEC MPEG and ITU-T VCEG (ISO/IEC JTC1/SC29/WG11 and ITU-T SG16 Q.6). Multiview video coding allows the construction of bit streams that represent multiple views [CHE200901]; MVC supports efficient encoding of video sequences captured simultaneously from multiple cameras using a single video stream. MVC can be used for encoding stereoscopic (two-view) and multiview 3DTV, and for free-viewpoint video (FVV) and free-viewpoint television (FVT).

MVC (ISO/IEC 14496-10:2008 Amendment 1 and ITU-T Recommendation H.264) is an extension of the AVC standard that provides efficient coding of multiview video. The encoder receives N temporally synchronized video streams and generates one bit stream. The decoder receives the bit stream, then decodes and outputs the N

video signals. Multiview video contains a large amount of inter-view statistical dependencies, since all cameras capture the same scene from different viewpoints. Therefore combined temporal and inter-view prediction is the key for efficient MVC. Also pictures of neighboring cameras can be used for efficient prediction [SMO200801]. MVC supports the direct coding of multiple views and exploits intercamera redundancy to reduce the bit rate. Although MVC is more efficient than simulcast, the rate of MVC-encoded video is proportional to the number of views.

The MVC group in the JVT has chosen the H.264/MPEG-4 AVC-based multiview video method as its MVC video reference model, since this method supports better coding efficiency than H.264/AVC simulcast coding. H.264/MPEG-4 AVC was developed jointly by ITU-T and ISO through the Joint Video Team (JVT) in the early 2000s (the ITU-T H.264 standard and the ISO/IEC MPEG-4 AVC, ISO/IEC 14496-10–MPEG-4 Part 10 are jointly maintained to retain identical technical content). H.264 is used with Blu-ray discs and videos from the iTunes store. The standardization of H.264/AVC was completed in 2003, but additional extensions have taken place since then, such as the scalable video coding as specified in Annex G of H.264/AVC added in 2007.

Appendix 4A: Additional Details on 3D Video Formats

This appendix expands on the 3D video formats that are available or under investigation. It applies to concepts covered in the body of the chapter more directly to home-based 3DTV applications. As might be expected, each approach has advantages and disadvantages as related to quality, efficiency (bit rate), complexity, and functionality.

4A.1 Conventional Stereo Video (CSV)

CSV is the most well developed and simplest 3D video representation. This approach only deals with color pixels of the video frames involved—no scene geometry processing is involved. The video signals are intended to be directly displayed using a 3D display system, although some video processing might take place after capture by the two (or perhaps more) cameras (for example, for normalization or color correction) and also before display. Figure 4A.1 shows an example of a stereo image pair; the same scene is visible from slightly different viewpoints. The 3D display system ensures that a viewer sees only the left view with the left eye and the right view with the right eye to create a 3D depth impression.

Compared to the other 3D video formats, the algorithms associated with CSV are the least complex; typically the algorithms act to separately encode and decode the multiple video signals, as shown in Figure 4A.2(a). The drawback is the fact that the amount of data is increasing compared to 2D video; however, reduction of resolution can be used to mitigate this requirement, as needed. It turns out that the MPEG-2 standard includes a MPEG-2 multiview profile coding

Figure 4A.1 (Please see color insert) Stereo image pair.

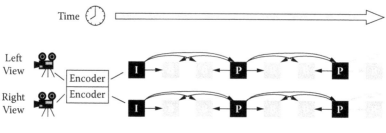

(a) Traditional MPEG-2/MPEG-4 applied to 3DTV

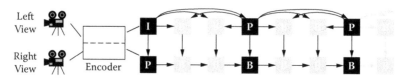

(b) MPEG-2 MultiView Profile and H.264/AVC SEI message

Figure 4A.2 (Please see color insert) Stereo video coding with combined temporal/inter-view prediction.

that allows efficiency to be increased by combining temporal/inter-view prediction as illustrated in Figure 4A.2(b).

As we noted in the body of the chapter, H.264/AVC was enhanced a few years back with a SEI message that can also be used to implement a prediction, as illustrated in Figure 4A.2(b). Although not designed for stereo-view video coding, the H.264 coding tools can be arranged to take advantage of the correlations between the pair of views of a stereo-view video and provide efficient compression. The scalable coding approach also alluded to in the body of the chapter can be supported by this extension since the requisite options can be signaled by the H.264 SEI messages; it is believed that these schemes could have coding performances similar to or better than that of proprietary system configurations for nonscalable coding [SUN200501].

For more than two views, the approach can be extended to multiview video coding (MVC) as illustrated in Figure 4A.3 [HUR200701]; MVC uses inter-view prediction by referring to the pictures obtained from the neighboring views. MVC has been standardized in the Joint Video Team (JVT) of the ITU-T Video Coding Experts Group (VCEG) and ISO/IEC MPEG. MVC enables efficient encoding of sequences captured simultaneously from multiple cameras using a single video stream. MVC is currently the most efficient way for stereo

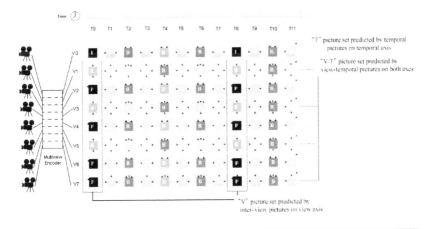

Figure 4A.3 (Please see color insert) Multiview video coding with combined temporal/inter-view prediction.

and multiview video coding; for two views the performance achieved by H.264/AVC stereo SEI message and MVC are similar [3DP200802]. MVC is also expected to become a new MPEG video coding standard for the realization of future video applications such as 3D Video (3DV) and free-viewpoint video (FVV). The MVC group in the JVT has chosen the H.264/AVC-based MVC method as the MVC reference model, since this method showed better coding efficiency than H.264/AVC simulcast coding and the other methods that were submitted in response to the call for proposals made by the MPEG [HUR200701], [SMO200401], [SUL200501], [MUE200601], [ISO200601].

A straightforward way to utilize existing video codecs (and infrastructure) for stereo video transmission is to apply one of the interleaving approaches illustrated in Figure 4A.4. A practical challenge is that there is no *de facto* industry standard available (so that any downstream decoder knows what kind of interleaving was used by the encoder). However, as noted elsewhere, there is a movement toward using an over/under approach.

Some new approaches are also emerging and have been proposed to improve efficiency, especially for bandwidth-limited environments. A new approach uses binocular suppression theory that employs disparate image quality in left- and right-eye views. Viewer tests have shown that (within reason) if one of the images of a stereopair is degraded, the perceived overall quality of the stereo video will be dominated by the higher-quality image [STE199801], [STE200001]. This concept is illustrated in

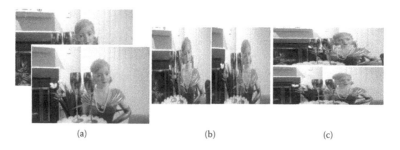

(a)   (b)   (c)

Figure 4A.4 (Please see color insert)   Stereo interleaving formats (a) time multiplexed frames, (b) spatial multiplexed as side-by-side, (c) spatial multiplexed as over/under.

Figure 4A.5 (Please see color insert)   Use of binocular suppression theory for more efficient coding.

Figure 4A.5. Applying this concept, one could code the right-eye image with less than the full resolution of the left eye—for example, downsampling it to half or quarter resolution (see Figure 4A.6). In principle this should provide comparable overall subjective stereo video quality while reducing the bit rate; if one were to adopt this approach, the 3D video functionality could be added by an overhead of say 25–30 percent to the 2D video for coding the right-eye view at quarter resolution.

The functionality of CSV is limited compared to the other 3D video formats described next. An intrinsic drawback of CSV is that the 3D impression cannot be modified: The baseline is fixed from capturing. Depth perception cannot be adjusted to different display types and sizes. The number of output views cannot be varied (only decreased). Head motion parallax cannot be supported (different perspective, occlusions, and disocclusions when moving the viewpoint) [3DP200802].

4A.2 Video plus Depth (V+D)

The concept of V+D representation is the next notch up in complexity. As can be seen in Figure 4A.7, a video signal and a per-pixel

Full resolution

Half resolution

Quarter resolution

Left eye   Right eye

Figure 4A.6 (Please see color insert)   Mixed resolution stereo video coding.

Video   Depth

Figure 4A.7 (Please see color insert)   Video plus depth (V+D) representation for 3D video.

depth map are captured and eventually transmitted to the viewer. The per-pixel depth data can be considered a monochromatic luminance signal with a restricted range spanning the interval $[Z_{near}, Z_{far}]$ representing, respectively, the minimum and maximum distance of the corresponding 3D point from the camera. The depth range is quantized with 8 bits, with the closest point having the value 255 and the most distant point having the value 0. Effectively the depth map is specified as a gray-scale image; these values can be supplied into the luminance channel of a video signal and the chrominance can be set to a constant value. In summary, this representation uses a regular video stream enriched with so-called depth maps providing a Z

value for each pixel. Note that V+D enjoys backward compatibility because a 2D receiver will display only the V portion of the V+D signal. Studies by the European ATTEST project (see Chapter 6) indicate that depth data can be compressed very efficiently and still get good quality, namely that it needs only around 20 percent of the bit rate that would otherwise be needed to encode the color video (these qualitative results were confirmed by means of subjective testing). This approach can be placed in the category of depth-enhanced stereo (DES).

A stereopair can be rendered from the V+D information by 3D warping at the decoder. A general warping algorithm takes a layer and deforms it in many ways: for example, twist it along any axis, bend a layer around itself, or add arbitrary dimension with a displacement map. Figure 4A.8 depicts the action of a general commercial warping program to illustrate the function. The generation of the stereopair from a V+D signal at the decoder is illustrated in Figure 4A.9. This reconstruction affords extended functionality compared to CSV because the stereo image can be adjusted and customized after transmission. Note that in principle more than two views can be generated at the decoder, thus enabling support of multiview displays (and head motion parallax viewing within reason).

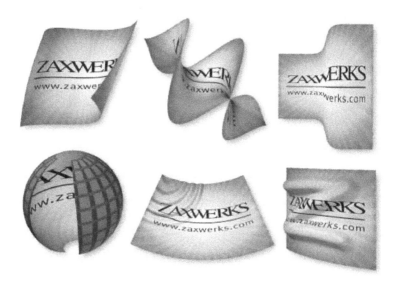

Figure 4A.8 (Please see color insert) Warping functions of a general commercial warping package. (Courtesy Zaxwerks, Inc. With permission.)

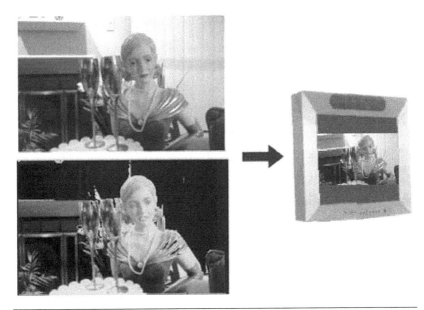

Figure 4A.9 (Please see color insert) Regeneration of stereo video from V+D signals.

V+D enjoys backward compatibility, compression efficiency, extended functionality, and the ability to use existing coding algorithms. It is only necessary to specify high-level syntax that allows a decoder to interpret two incoming video streams correctly as color and depth. The specifications "ISO/IEC 23002–3 Representation of Auxiliary Video and Supplemental Information" and "ISO/IEC 13818–1:2003 Carriage of Auxiliary Data" enable 3D video-based V+D to be deployed in a standardized fashion by broadcasters interested in going this way.

It should be noted, however, that the advantages of V+D over CSV entail increased complexity for both the sender's side and the receiver's side. At the receiver side, view synthesis has to be performed after decoding to generate the second view of the stereopair. At the sender (capture) side, the depth data have to be generated before encoding can take place. This is usually done by depth/disparity estimation from a captured stereopair; these algorithms are complex and still error prone. Thus, in the near future, V+D might be more suitable for applications with playback functionality, where depth estimation can be performed offline on powerful machines, such as in a production studio or home 3D-editing suite, enabling viewing of downloaded 3D video clips and 3DTV broadcasting [3DP200802].

4A.3 Multiview Video plus Depth (MV+D)

There are some advanced 3D video applications that are not properly supported by any existing standards and where work by the ITU-R or ISO/MPEG is needed. Two such applications are

- Wide-range multiview autostereoscopic displays (say, nine or more views)
- Free-viewpoint video (environment where the user can choose his or her own viewpoint)

These 3D video applications require a 3D video format that allows rendering a continuum or a large number of output views at the decoder. There really are no available alternatives: MVC discussed above does not support a continuum and becomes inefficient for a large number of views; and we noted that V+D could in principle generate more than two views at the decoder, but in practicality it supports only a limited continuum around the original view (artifacts increase significantly with distance of the virtual viewpoint). In response to these requirements MPEG started an activity to develop a new 3D video standard that would support these requirements.

The MV+D concept is illustrated in Figure 4A.10. MV+D involves a number of complex processing steps where depth has to be estimated for the N views at the capture point, and then N color with

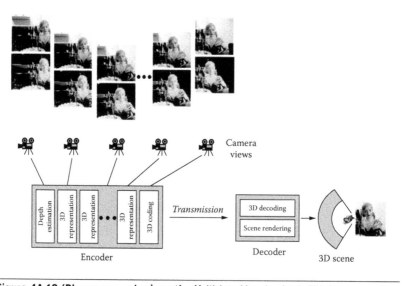

Figure 4A.10 (Please see color insert)   Multiview video plus depth (MV+D) concept.

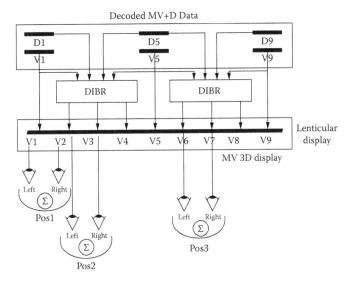

Figure 4A.11   Multiview autostereoscopic displays based on MV+D.

N-depth videos streams have to be encoded and transmitted. At the receiver, the data have to be decoded and the virtual views have to be rendered (reconstructed).

MV+D can be used to support multiview autostereoscopic displays in a relatively efficient manner. Consider a display that supports nine views (V1–V9) simultaneously (for example, with a lenticular display manufactured by Philips) (see Figure 4A.11). From a specific position a viewer can see only a stereopair of views, depending on the viewer's position. Transmitting nine display views directly (for example, by using MVC) would be taxing from a bandwidth perspective; in this illustrative example only three original views (views V1, V5, and V9) along with corresponding depth maps D1, D5, and D9 are in the decoded stream; the remaining views can be synthesized from these decoded data by using depth-image-based rendering (DIBR) techniques.

4A.4 Layered Depth Video (LDV)

LDV is a derivative and also an alternative to MV+D. LDV is believed to be more efficient than MV+D because less information has to be transmitted; however, additional error-prone vision-processing tasks are required that operate on partially unreliable depth data, and these efficiency assessments remain fully valid as of press time.

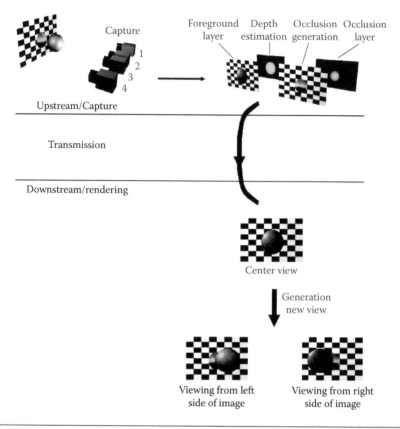

Figure 4A.12 (Please see color insert)   Layered depth video (LDV) Concept.

LDV uses (1) one color video with associated depth map and (2) a background layer with associated depth map; the background layer includes image content that is covered by foreground objects in the main layer. This is illustrated in Figures 4A.12 and 4A.13. The occlusion information is constructed by warping two or more neighboring video-plus-depth views from the MV+D representation onto a defined center view. The LDV stream or substreams can then be encoded by a suitable LDV coding profile.

Note that LDV can be generated from MV+D by warping the main layer image onto other contributing input images (such as an additional left and right view). By subtraction it is then determined which parts of the other contributing input images are covered in the main layer image; these are then assigned as residual images and transmitted while the rest is omitted [3DP200802].

Capture

Foreground Layer

Depth Estimation

Occlusion Generation

Occlusion Layer

Figure 4A.13 (Please see color insert)   Layered depth video (LDV) example. (A. Frick, F. Kellner, B. Bartczak, and R. Koch: Generation of 3D-TV LDV-Content with Time of Flight Camera. Proceedings of 3DTV-CON 2009, Potsdam May 04–06, 2009.)

Figure 4A.14 is based on a recent presentation at the 3D Media Workshop, HHI Berlin, October 15–16, 2009 [KOC200901], [FRI200901]. LDV provides a single view with depth and occlusion information. The goal is to achieve automatic acquisition of 3DTV content, especially how to obtain depth and occlusion information from video and how to extrapolate new views without error. Figure 4A.15, also from the same workshop, depicts how an end-to-end system for LDV can be assembled.

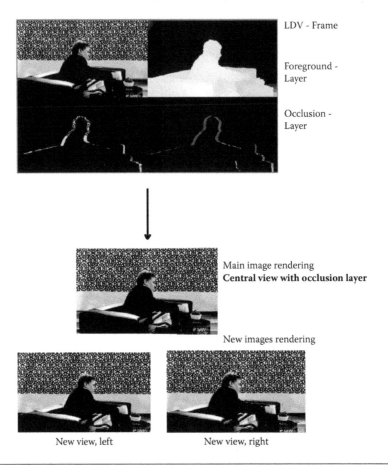

Figure 4A.14 (Please see color insert)   Another view of LDV.

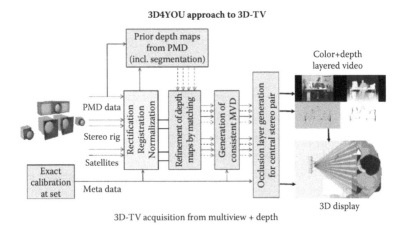

Figure 4A.15   End-to-end system for LDV in the 3D4YOU European project.

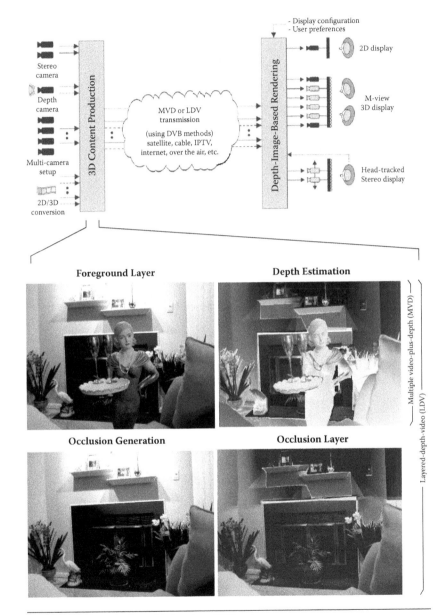

Figure 1.7 Example of possible long-term 3DTV system.

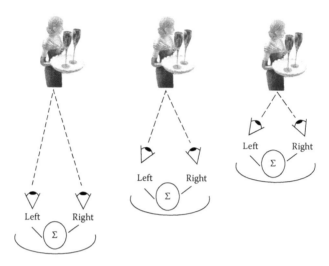

Figure 2.2 Convergence.

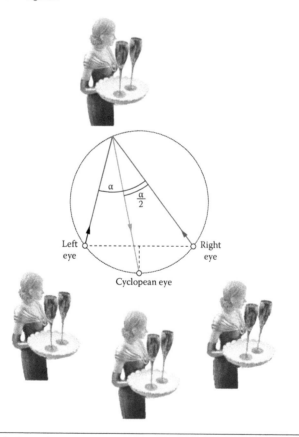

Figure 2.3 Cyclopean image.

Figure 3.1 Stereoscopic capture of scene to achieve 3D when scene is seen with appropriate display system.

Figure 3.11 HPLabs-UCBerkeley Multiview Facility: (left) projectors; (right) display and cameras.

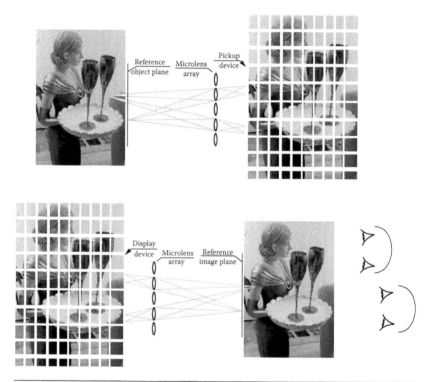

Figure 3.13 Integral imaging principles.

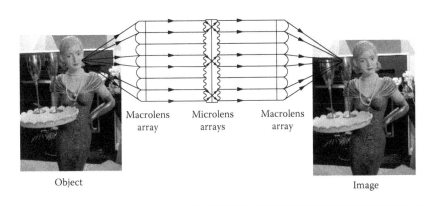

Macrolens array · Microlens arrays · Macrolens array

Object

Image

Figure 3.14 An approach to holoscopic imaging.

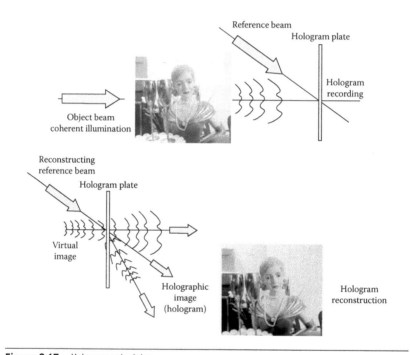

Figure 3.15 Hologram principles.

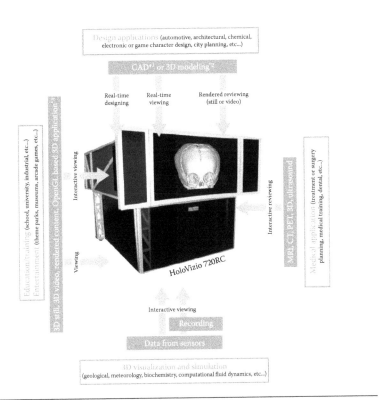

Design applications (automotive, architectural, chemical, electronic or game character design, city planning, etc...)

CAD[*1] or 3D modeling[*2]

Real-time designing

Real-time viewing

Rendered reviewing (still or video)

Interactive viewing

Education/Training (school, university, industrial, etc...)

Entertainment (theme parks, museums, arcade games, etc...)

3D still, 3D video, rendered content, OpenGL based 3D application[*3]

Viewing

HoloVizio 720RC

Interactive reviewing

MRI, CT, PET, 3D, ultrasound

Medical application (treatment or surgery planning, medical training, dental, etc...)

Interactive viewing

Recording

Data from sensors

3D visualization and simulation (geological, meteorology, biochemistry, computational fluid dynamics, etc...)

Figure 3.16 Holografika hybrid system (high-end applications).

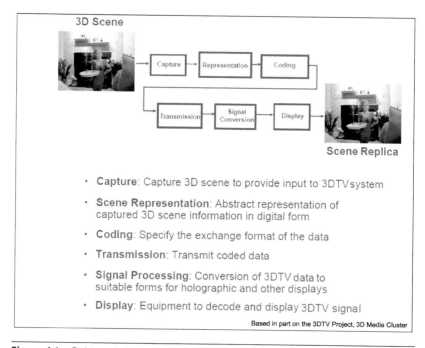

- **Capture**: Capture 3D scene to provide input to 3DTV system
- **Scene Representation**: Abstract representation of captured 3D scene information in digital form
- **Coding**: Specify the exchange format of the data
- **Transmission**: Transmit coded data
- **Signal Processing**: Conversion of 3DTV data to suitable forms for holographic and other displays
- **Display**: Equipment to decode and display 3DTV signal

Based in part on the 3DTV Project, 3D Media Cluster

Figure 4.1 End-to-end signal management—general view.

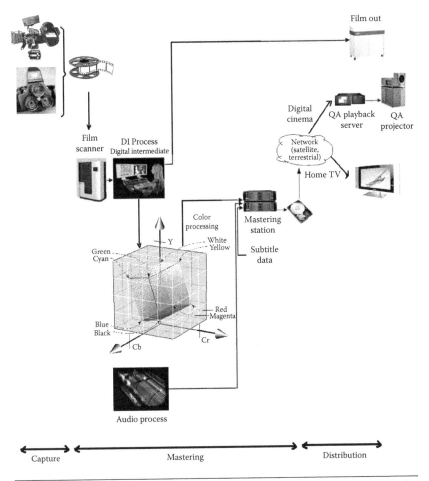

Film out

Digital cinema

QA playback server

QA projector

Network (satellite, terrestrial)

Home TV

Film scanner

DI Process
Digital intermediate

Color processing

Mastering station

Subtitle data

Y
White
Yellow
Green
Cyan
Red
Magenta
Blue
Black
Cb
Cr

Audio process

Capture   Mastering   Distribution

Figure   4.2 Overall   video   capture/mastering/distribution   process—implementation perspective.

Figure 4.4 Spatial compression.

Figure 4.10 Anaglyth method—another view.

Figure 4A.1 Stereo image pair.

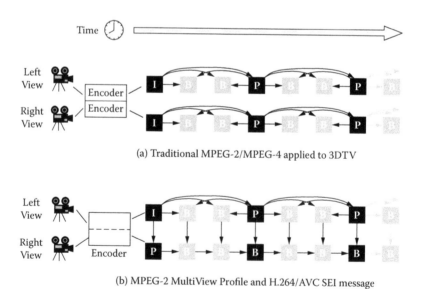

(a) Traditional MPEG-2/MPEG-4 applied to 3DTV

(b) MPEG-2 MultiView Profile and H.264/AVC SEI message

Figure 4A.2 Stereo video coding with combined temporal/inter-view prediction.

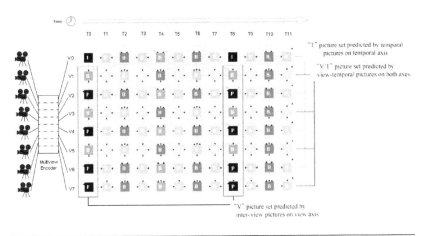

Figure 4A.3 Multiview video coding with combined temporal/inter-view prediction.

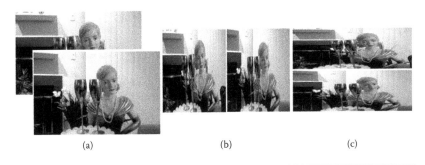

(a)   (b)   (c)

Figure 4A.4 Stereo interleaving formats (a) time multiplexed frames, (b) spatial multiplexed as side-by-side, (c) spatial multiplexed as over/under.

Figure 4A.5 Use of binocular suppression theory for more efficient coding.

Full resolution

Half resolution

Quarter resolution

Left eye   Right eye

Figure 4A.6 Mixed resolution stereo video coding.

Video   Depth

Figure 4A.7   Video plus depth (V+D) representation for 3D video.

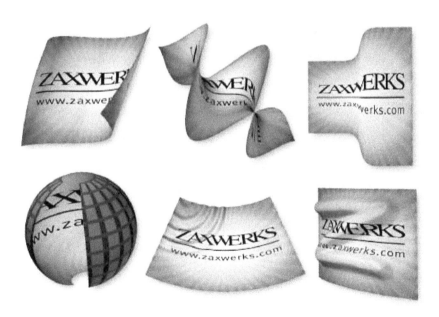

Figure 4A.8   Warping functions of a general commercial warping package.

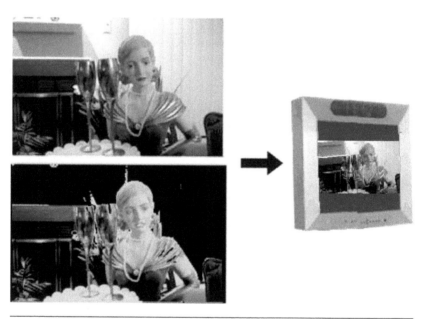

Figure 4A.9   Regeneration of stereo video from V+D signals.

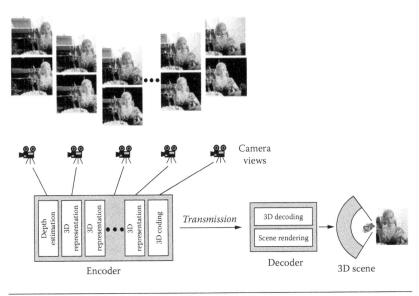

Figure 4A.10 Multiview video plus depth (MV+D) concept.

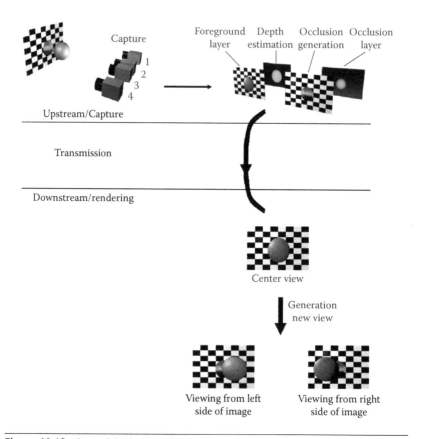

Figure 4A.12 Layered depth video (LDV) Concept.

Capture

Foreground Layer

Depth Estimation

Occlusion Generation

Occlusion Layer

Figure 4A.13   Layered depth video (LDV) example.

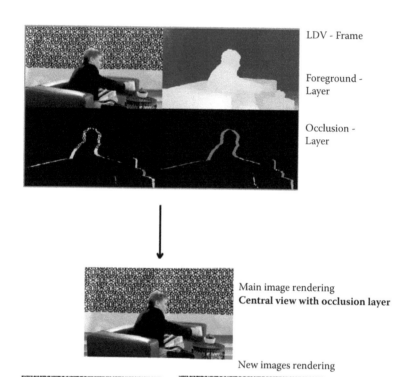

LDV - Frame

Foreground - Layer

Occlusion - Layer

Main image rendering
Central view with occlusion layer

New images rendering

New view, left   New view, right

Figure 4A.14 Another view of LDV.

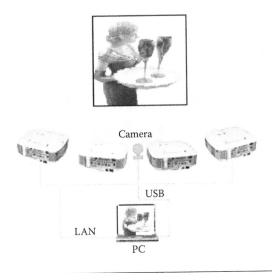

Figure 5.3 Stacked (front) projection concept.

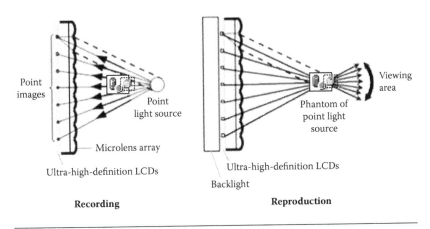

Figure 5.7 Principle of integral photography.

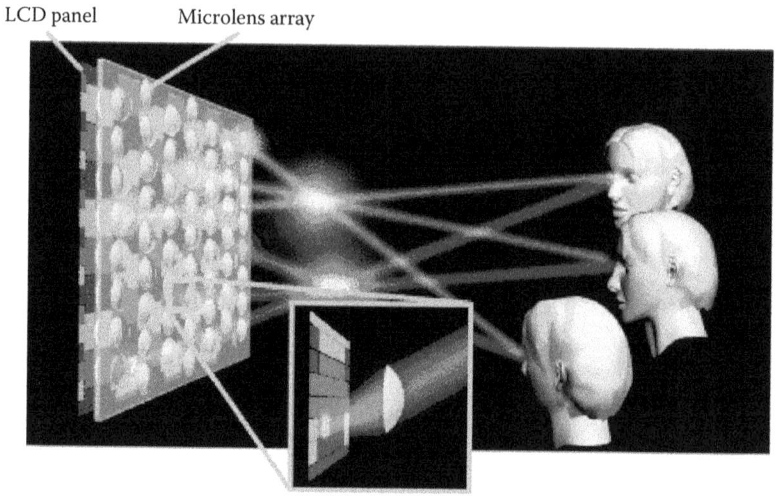

LCD panel   Microlens array

Optimum color-filter arrangement for realizing light-ray-direction
multiplication and moire reduction

Figure 5.8   Display technology to support integral videography.

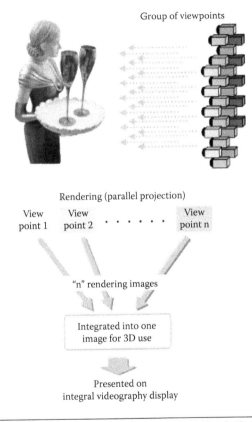

Group of viewpoints

Rendering (parallel projection)

View point 1   View point 2   · · · · · ·   View point n

"n" rendering images

Integrated into one image for 3D use

Presented on integral videography display

Figure 5.9   Hitachi prototype "integral photography with overlaid projection" concept.

Figure 6.1   3DPresence multiparty videoconferencing concept.

Figure 6.2   Drawing of the mock-up system.

References

[3DP200802] 3DPHONE, Project no. FP7–213349, Project title ALL 3D Imaging Phone, 7th Framework Programme, Specific Programme "Cooperation," FP7-ICT-2007.1.5—Networked Media, D5.1—Requirements and specifications for 3D video, 19 August 2008.

[3DT200801] IST–6th Framework Programme, 3DTV NoE, 2004, Project Coordinator Prof. L. Onural, EEE Department, Bilkent University, TR-06800 Ankara, Turkey.

[CHE200701] Y. Chen, M.D. Adams, and W.S. Lu, Design of optimal quincunx filter banks for image coding, *EURASIP J. Adv. Signal Process.*, 2007.

[CHE200901] Y. Chen, Y.-K. Wang, K. Ugur, M.M. Hannuksela, J. Lainema, and M. Gabbouj, The emerging MVC standard for 3D video services, *EURASIP J. Adv. Signal Process.*, 2009.

[D32200701] D32.2, Technical report #2 on 3D telecommunication issues, Project Number 511568, Project Acronym 3DTV, *Integrated Three-Dimensional Television—Capture, Transmission and Display*, 20 February 2007, Murat Tekalp, Ed.

[DOS200801] C. Dosch and D. Wood, Can we create the "holodeck"? The challenge of 3D television, *ITU News Magazine*, no. 9, November 2008.

[EDC200701] The European Digital Cinema Forum (EDCF), *The EDCF Guide to Digital Cinema Mastering*, Somerset, UK, EDCF, August 2007.

[FEH200201] C. Fehn, P. Kauff, et al., An evolutionary and optimized approach on 3DTV, *Proceedings of International Broadcast Conference '02*, Amsterdam, 357–365, 2002.

[FEH200301] C. Fehn, A 3DTV approach using depth-image-based rendering (DIBR), *Proceedings of Visualization, Imaging, and Image Processing '03*, Benalmadena, Spain, 482–487, 2003.

[FEH200601] C. Fehn, Depth-Image-Based Rendering (DIBR), Compression, and Transmission for a Flexible Approach on 3DTV, Ph.D. thesis, Technical University Berlin, Germany, 2006.

[FRI200901] A. Frick, F. Kellner, et al., Generation of 3DTV LDV content with time of flight cameras, *Proceedings of 3DTV-CON 2009*, Potsdam, 4–6 May 2009.

[GOU200001] A. Gouze, M. Antonini, and M. Barlaud, Quincunx lifting scheme for lossy image compression, in *Proceedings of the IEEE International Conference on Image Processing*, 1, 665–668, September 2000.

[HUR200701] J-H. Hur, S. Cho, and Y-L. Lee, Illumination change compensation method for H.264/AVC-based multiview video coding, *IEEE Trans. Circuits Syst. Video Technol.*, 17, 11, November 2007.

[IMD201001] Internet Movie Database, Glossary, http://www.imdb.com

[ISO200601] ISO, Subjective Test Results for the CfP on Multi-View Video Coding, ISO/IEC JTC1/SC29/WG11, Bangkok, Thailand, Doc. N7779, 2006.

[ISO200901] International Organization for Standardization, ISO/IEC JTC1/SC29/WG11, Coding of Moving Pictures and Audio, Vision on 3D Video, Video and Requirements, ISO/IEC JTC1/SC29/WG11 N10357, Lausanne, Switzerland, February 2009.

[JOH200901] C. Johnston, Will new year of 3D drive lens technology? *TV Technology Online Magazine*, 15 December 2009.

[KOC200901] R. Koch, Future 3DTV acquisition, 3D Media Workshop, HHI Berlin, 15–16 October 2009.

[KOV200001] J. Kovacevic and W. Sweldens, Wavelet families of increasing order in arbitrary dimensions, *IEEE Trans. Image Process.*, 9, 3, 480–496, March 2000.

[LIU200601] Y. Liu, T.T. Nguyen, and S. Oraintara, Embedded image coding using quincunx directional filter bank, *ISCAS 2006*, IEEE, 4943 ff.

[MER200901] R. Merritt, Incomplete 3DTV products in CES spotlight HDMI upgrade one of latest pieces in stereo 3D puzzle, EE *Times*, 23 December 2009.

[MIN199401] D. Minoli and R. Keinath, *Distributed Multimedia through Broadband Communication Services*, Artech House, Boston, 1994.

[MIN199501] D. Minoli, *Video Dialtone Technology: Digital Video over ADSL, HFC, FTTC, and ATM*, McGraw-Hill, New York, 1995.

[MIN199601] D. Minoli, *Distance Learning: Technology and Applications*, Artech House, Boston, 1996.

[MIN200001] D. Minoli, Digital video, in *The Telecommunications Handbook*, K. Terplan and P. Morreale, Eds., CRC Press, Boca Raton, FL, 2000.

[MIN200801] D. Minoli, *IP Multicast with Applications to IPTV and Mobile DVB-H*, Wiley/IEEE Press, Hoboken, NJ, 2008.

[MIN200802] J. Amoss and D. Minoli, *Handbook of IPv4 to IPv6 Transition: Methodologies for Institutional and Corporate Networks*, Auerbach Publications, Boca Raton, FL, 2008.

[MIN200901] D. Minoli, *Satellite Systems Engineering in an IPv6 Environment*, CRC Press, Boca Raton, FL, 2009.

[MUE200601] K. Mueller, P. Merkle, A. Smolic, and T. Wiegand, Multiview coding using AVC, ISO/IEC JTC1/SC29/WG11, Bangkok, Thailand, Doc. M12945, 2006.

[ONU200601] L. Onural, A. Smolic, et al., An assessment Of 3DTV technologies, *2006 NAB BEC Proceedings*, 456ff.

[SMO200401] A. Smolic and D. McCutchen, 3DAV exploration of video-based rendering technology in MPEG, *IEEE Trans. Circuits Syst. Video Technol.*, 14, 3, 348–356, March 2004.

[SMO200801] A. Smolic, Ed., Introduction to multiview video coding, January 2008, Antalya, Turkey, International Organization for Standardization, ISO/IEC JTC 1/SC 29/WG 11, Coding of moving pictures and audio.

[STA200901] M. Starks, Spacespex™ anaglyph—The only way to bring 3DTV to the masses, Online article, 2009.

[STE199801] L. Stelmach and W.J. Tam, Stereoscopic image coding: effect of disparate image-quality in left- and right-eye views, *Signal Process. Image Comm.*, 14, 111–117, 1998.

[STE200001] L. Stelmach, W.J. Tam, D. Meegan, and A. Vincent, Stereo image quality: Effects of mixed spatio-temporal resolution, *IEEE Trans. Circuits Syst. Video Technol.*, 10, 2, 188–193, March 2000.

[SUL200501] G. Sullivan, T. Wiegand, and A. Luthra, Draft of Version 4 of H.264/AVC (ITU-T Recommendations H.264 and ISO/IEC 14496–10 (MPEG-4 Part 10) Advanced Video Coding), ISO/IEC JTC1/SC29/WG11 and ITU-T Q6/SG16, 2005.

[SUN200501] S. Sun and S. Lei, Stereo-view video coding using H.264 tools, *Proc. SPIE—Int. Soc. Opt. Eng.*, 5685 177–184, 2005.

[SWE199601] W. Sweldens, The lifting scheme: A custom-design construction of biorthogonal wavelets, *Appl. Comput. Harmonic Anal.*, 3, 186–200, 1996.

[TAY199301] D.B.H. Tay and N.G. Kingsbury, Flexible design of multidimensional perfect reconstruction FIR 2-band filters using transformations of variables, *IEEE Trans. Image Process.*, 2, 4, 466–480, October 1993.

[ZHA200701] X. Zhang, X. Wu, and F. Wu, Image coding on quincunx lattice with adaptive lifting and interpolation, 2007 Data Compression Conference (DCC'07), IEEE Computer Society.

5

3D Basic 3DTV Approaches and Technologies for In-Home Display of Content

In our discussion so far we have been tracing a 3D signal from mastering to home delivery. We noted in previous chapters that 3D content may be available at home from a Blu-ray disk (BD) player, a PC, or set-top box (this being fed from a terrestrial cable TV system, a satellite DTH link, or eventually also from terrestrial over-the-air broadcasting). The next step in the end-to-end delivery of the content is to be able to display it on an appropriate screen, as depicted in Figure 5.1. The display in a viewer's home is the most obvious and conspicuous element of the 3DTV system, besides the ultimate content. At press time several TV makers, including Sony, Sharp, Vizio, and Panasonic, announced plans to sell 3D televisions to consumers in 2010. The technologies being considered for 3D display can be broadly grouped into the following categories, listed here in likely order of general consumer acceptance (acceptance being based on quality and cost): stereoscopic displays, autostereoscopic displays, head mounted displays,* volumetric displays, and holographic displays. Stereoscopic systems require active or passive glasses but are likely to see the earliest home penetration. This chapter examines some of the issues involved with the home portion of the end-to-end path. We have discussed various aspects of display technology throughout this text, and hence we focus mostly on technology delivery (real, purchasable systems) in this chapter.

---

* Head mounted displays are not covered in this text.

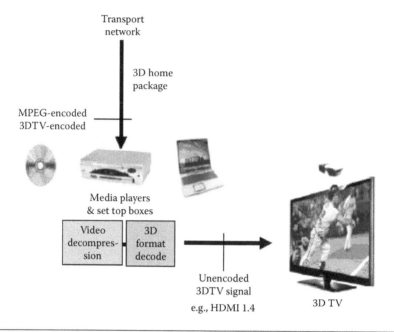

Figure 5.1 In-home 3DTV signal management.

5.1 Connecting the In-Home Source to the Display

For the purpose of this discussion, at this juncture the 3DTV content is ready to be displayed on the screen over an uncompressed link. A number of in-home physical connectivity arrangements are possible, including DVI-D, DVI-I, HDMI, and DisplayPort, or even some wireless connections such as IEEE 802.11, Wireless Gigabit Alliance, or a proprietary system such as WirelessHD. (See Appendix 5A for a description of these various in-house connectivity systems.) In the short term this physical connection will likely be HDMI;* TV displays available at press time typically support HDMI 1.3, but an upgrade will be required (at both ends of the unencoded-signal cable).

A current question facing developers is where the decoding of the 3DTV signal will be done in the home. At press time the Consumer Electronics Association (CEA) had just developed standards for the interface for an uncompressed digital interface between, say, the set top box (called *source*) and the 3D display (called *sink*); these standards

---

* In some instances two cables are needed to operate two twin projectors; in most instances, however, a single cable is used.

will need to include signaling details, 3D format support, and other interoperability requirements between sources and sinks [CHI200901]. This entailed an upgrade of the CEA 861 standard (A DTV Profile for Uncompressed High Speed Digital Interfaces, March 2008) that defines an uncompressed video interconnect for HDMI. On HDMI 1.4 cables, 3D content can travel across the cable interface, with or without coding. The decoding could be handled in the BD player, at the set-top box, or even within a PC (if that is the display terminal). Alternatively, the decoding could be handled by the TV/display device. As a minimum, some transcoding from the incoming 3D stream (hopefully in a standardized format) will be needed at the display device to support the native (internal) 3D mechanism of the specific display device. The 1.4 upgrade of the HDMI interface will pave a way for future stereo 3D broadcasts to be transported locally on new and existing HDMI cables; however, as noted, new chipsets will be needed at both ends of that interface to make full use of the new features, including the 3DTV features.

In mid 2009 an HDMI 1.4 specification was announced. It was positioned as follows:

> The HDMI specification continues to add functionality as the consumer electronics and PC industries build products that enhance the consumer's HD experience. The [HDMI] 1.4 specification will support some of the most exciting and powerful near-term innovations such as Ethernet connectivity and 3D formats. Additionally we are going to broaden our solution by providing a smaller connector for portable devices. [COL200901]

TV displays being brought to the market in 2010 were expected to support the new 1.4 specification that, as noted, adds optional support for 3D signals. The expectation is also that the firmware in some higher-end BD players can be updated to support the 3D signaling methods included in HDMI 1.4.

These new mechanisms come close to being able to support a full 3D experience, but with some compromises in the short term. HDMI 1.3 supports 2D displays at 1080p/50 or 60 Hz; 148 MHz silicon is used in the drives. HDMI 1.4 reallocates this bandwidth to allow 3D support for 1080p/24, 720p/50, or 720p/60 per eye (dual stream). This is adequate for a 720p HD image in each eye, but not to support a full 1080p HD image per eye, namely 1080p/60 [CHI200901].* The expectation is that BD players with HDMI 1.4 interfaces will be able to output 1080p/24 signals to the TV display; this signal will have to be "frame-rate converted" to display content at 120 Hz per eye. See Appendix 5A for a more detailed discussion.

Related to this, the Blu-ray Disc Association (BDA) announced the finalization and release of the Blu-ray 3D specification at the end of 2009. The specification, which represents work undertaken by the leading Hollywood studios and consumer electronic and computer manufacturers, will enable the home entertainment industry to bring the stereoscopic 3D experience into consumers' living rooms on BD but will require consumers to acquire new players, high-definition TVs, and shutter glasses. The Blu-ray 3D specification encodes 3D video using the multiview video coding (MVC) codec, an extension to the ITU-T H.264 advanced video coding (AVC) codec currently supported by all BD players, as we alluded to in the previous chapter. MPEG4-MVC compresses both left- and right-eye views with a typical 50 percent overhead compared to equivalent 2D content, according to BDA and can provide full 1080p resolution backward compatibility with current 2D BD players [SHI200901].

5.2 3DTV Display Technology

Now that the signal can be brought to the TV set, it needs to be displayed in a compelling and attractive manner. To that end, a variety of displays are being developed for 3DTV home use. Current 3DTV displays are based on 2D display methods with some technology extensions to allow them to render two views (one for each eye) temporally or spatially interleaved; future displays may be based on different technologies and designs.

We first review briefly some terms introduced in Chapter 2. *Binocular parallax* is the difference in the appearance of an object as seen by the right and left eyes; binocular parallax makes depth perception possible. Binocular parallax arises from the difference in

---

* To achieve this higher performance new silicon drivers operating at 297 MHz are needed; HDMI may likely support this rate in the future, but it was not part of HDMI 1.4.

position that two objects of different distances are projected onto the retinas of the left and right eyes. *Full parallax* refers to the ability to select a viewpoint by the viewer at image/video display time, not just the viewpoint offered by the image-capturing camera itself. Any 3D display screen needs to generate parallax, which in turn creates a stereoscopic sense. A 3D video image that can be viewed in full parallax is called "full parallax" video; this typically requires holographic or volumetric techniques, and since height, orientation, and distance of the viewpoint in relation to the subject can be adjusted, the user can view the subject from the angle he or she desires and the scene can be watched regardless of the viewing direction (that is, from above, below, left, or right) [OIK200801]. The ultimate goal is to be able to deploy full parallax displays for 3DTV viewing in the home, but this will likely not happen for a number of years.

Figure 5.2 provides a taxonomy of 3D display technologies. Four main categories are noted in the figure: stereoscopic (also known as "two-view screen"), autostereoscopic, volumetric, and holographic (also known as light field). The figure also includes a sense of timing, cost, and applications of the technology. Many of the displays currently on the market or just being brought to market are for stereoscopic viewing with glasses; all stereoscopic systems have one thing in common: the need to independently present left- and right-eye images to the corresponding eye. Other available displays are autostereoscopic. Some displays support multiple views of a scene (supporting horizontal parallax). The long-term goal is to have a system that does not require glasses and supports both horizontal and vertical parallax, but that technology may require a number of years to develop for home use.

The form that such displays would take is one aspect that needs considerable thought and is a major concern in consumer acceptance. The main requirement of a 3D display is to create the illusion of depth or distance by using a series of depth cues such as disparity, motion parallax, and ocular accommodation. Conflicting cues are one of the leading causes for discomfort and fatigue when viewing 3D displays. Important aspects to be considered include image resolution, field of view, brightness, number of users (single or multiple), viewing distance, and cost [ONU200601].

A number of advancements have been achieved in recent years relating to TV displays in general and 3DTV displays in particular, including wider visual field angles, display of an expanded range of colors,

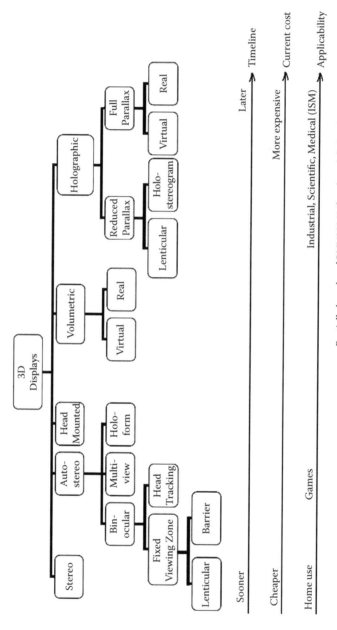

Figure 5.2 Taxonomy of 3D display technology.

higher contrast, higher definition in animation displays, reduction in size and weight, very low power consumption, ultra-slim bodies, larger screens, larger capacity for higher efficiency, and more advanced functionality [MAY200901]. Table 5.1 provides a glossary of key display-related concepts and terms, based partially on reference [KIN200901].

Stereoscopic displays can be of the "passive stereo" category or "active stereo" category. Passive stereo refers to the glasses being used, namely with glasses that have no electronic components. Active stereo systems use glasses with liquid crystal display (LCD) shutters and electronics that change the state of the LCD from transparent to opaque very rapidly and in synchrony with the projection of the stereo pairs, as discussed in Chapter 3. It is the expectation of the industry that in the first phase of 3DTV deployment in the home, the consumer will use stereoscopic displays with glasses (autostereoscopic, volumetric, and light field will follow later in the decade, at various points in time—it could be 10 years before high-quality, reasonably priced commercial off-the-shelf displays based on these technologies are available to the home consumer [CHI200901]).

At the constituent technology level, screens can use projection, LCD (liquid crystal display), DLP (digital light processing), organic light-emitting diode (OLED), or other approaches. Each technology has certain advantages over the other, and each has limitations. A brief description of each of these technologies follows [KIN200901], [POW200901]:

- DLP is a projection TV technology developed by Texas Instruments, based on their digital micromirror device (DMD) microchip. Each DMD chip has an array of tiny swiveling mirrors that create the image. Depending on the TV's resolution, the number of mirrors can range from several hundred thousand to over two million. In a DLP projector, light from the projector's lamp is directed onto the surface of the DLP chip. The mirrors tilt back and forth, directing light either into the lens path to turn the pixel on or away from the lens path to turn it off. DLP technology is used in both front- and rear-projection displays. There are two basic types of DLP projectors. "Single-chip" models (virtually all rear-projection DLP TVs) use a single DMD chip, with color provided by a spinning color wheel or colored LEDs. "Three-

Table 5.1 Glossary of Key Display-Related Terms

1080p	1080p is a high-definition video format with resolution of 1920×1080 pixels. The "p" refers to "progressive scan," which means that each video frame is transmitted as a whole in a single sweep. The main advantage of 1080p TVs is that they can display all high-definition video formats without down-converting, which sacrifices some picture detail. 1080p TVs display video at 60 frames per second, so this format is often referred to as 1080p60. The video on most high-definition discs is encoded at the film's native rate of 24 frames per second, or 1080p24. For compatibility with most current 1080p TVs, high-definition players internally convert the 1080p24 video to 1080p60. By 2008, many HDTVs included the ability to accept a 1080p24 signal directly. These TVs do not actually display video at 24 frames per second because that would cause visible flicker and motion stutter. The TV converts the video to 60 frames per second or whatever its native display rate is. The ideal situation would be to display 1080p24 at a multiple of 24 frames per second, such as 72, 96, or 120 frames per second, to avoid the motion judder caused by the 3–2 pull-down that is required when converting 24-frames-per-second material to 60 frames per second.
120 Hz refresh rate	The digital display technologies (LCD, plasma, DLP, LCoS, etc.) that have replaced picture tubes progressive scan by nature, displaying 60 video frames per second—often referred to as "60 Hz." HDTVs with 120 Hz refresh rate employ sophisticated video processing to double the standard rate to 120 frames per second by inserting either additional video frames or black frames. Because each video frame appears for only half the normal amount of time, on-screen motion looks smoother and more fluid, with less smearing. It is especially noticeable viewing fast-action sports and video games. This feature is available on an increasing number of flat-panel LCD TVs.
16:9	See aspect ratio and widescreen.
240 Hz refresh rate	240 Hz refresh rate reduces LCD motion blur even more than 120 Hz refresh rate. 240 Hz processing creates and inserts three new video frames for every original frame. Most "240 Hz" TVs operate this way, but some models use "pseudo-240 Hz" technology that combines 120 Hz refresh rate with high-speed backlight scanning. An example of a pseudo-240 Hz approach that is very effective is Toshiba's ClearScan 240 technology.
ALiS (Alternate Lighting of Surfaces)	A type of high-definition plasma TV panel designed for optimum performance when displaying 1080i material. On a typical progressive-scan plasma TV, all pixels can be illuminated at any given instant. With an ALiS plasma panel, alternate rows of pixels are illuminated so that only half the panel's pixels can be illuminated at any moment (somewhat similar to interlaced scanning on a CRT-type TV). ALiS-based plasmas make up a small part of the overall market; TV makers that use ALiS panels include Hitachi and Fujitsu.
Anamorphic video	Refers to widescreen video images that have been "squeezed" to fit a narrower video frame when stored on DVD. These images must be expanded (unsqueezed) by the display device. Most of today's TVs employ a screen with 16:9 aspect ratio so that anamorphic and other widescreen material can be viewed in its proper proportions. When anamorphic video is displayed on an old-fashioned TV with a 4:3 screen, images appear unnaturally tall and narrow.

Table 5.1 (*Continued*)   Glossary of Key Display-Related Terms

Aspect ratio	The ratio of width to height for an image or screen. The North American National Television System Committee (NTSC) television standard uses the squarish 4:3 (1.33:1) ratio. HDTVs use the wider 16:9 ratio (1.78:1) to better display widescreen material like high-definition broadcasts and DVDs.

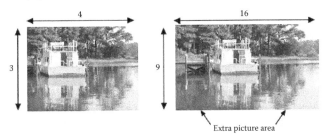

Backlight scanning	An anti-blur technology used in some LCD TVs. Typical LCDs use a fluorescent backlight that shines constantly, which can contribute to motion blur. LCD models with backlight scanning have a special type of fluorescent backlight that pulses at very high speed, which has the effect of reducing motion blur. Some recent TVs use backlight scanning along with 120 Hz refresh rate for even greater blur reduction.
Contrast ratio	Measures the difference between the brightest whites and the darkest blacks that a TV can display. The higher the contrast ratio, the better a TV will be at showing subtle color details, and the better it will look in rooms with more ambient room light. Contrast ratio is one of the most important specs for all TV types. There are two ways of measuring a TV's contrast ratio. Static contrast ratio measures the difference between the brightest and darkest images a TV can produce simultaneously (sometimes called on-screen contrast ratio). The ratio of the brightest and darkest images a TV can produce over time is called dynamic contrast ratio. Both specs are meaningful, but the dynamic spec is often four or five times higher than the static spec.
Direct-view TV	A general term for nonprojection types of TVs, which include conventional tube TVs and flat-panel plasma and LCD TVs.
DLNA (Digital Living Network Alliance)	A collaborative initiative among more than 200 companies, including Sony, Panasonic, Samsung, Microsoft, Cisco, Denon, and Yamaha. Their goal is to create products that connect to each other across one's home network, regardless of manufacturer, so one can easily enjoy one's digital and online content in any room. While all DLNA-compliant devices are essentially guaranteed to work together, they may not be able to share all types of media. For example, a DLNA-certified TV might be able to display digital photos from a DLNA-certified media server, but not videos.
DLP (digital light processing)	A projection TV technology developed by Texas Instruments based on their digital micromirror device (DMD) microchip. Each DMD chip has an array of tiny swiveling mirrors that create the image. Depending on the TV's resolution, the number of mirrors can range from several hundred thousand to over two million.

Continued

Table 5.1 (*Continued*)   Glossary of Key Display-Related Terms

DVI (digital visual interface)	A multipin, computer-style connection intended to carry high-resolution video signals from video source components (such as older HD-capable satellite and cable boxes, and up-converting DVD players) to HD-capable TVs with a compatible connector. Most (but not all) DVI connections use HDCP (high-bandwidth digital content protection) encryption to prevent piracy. In consumer electronics products, DVI connectors have been almost completely replaced by HDMI connectors that carry both video and audio. One can use an adapter to connect a DVI-equipped component to an HDMI-equipped TV, or vice versa, but a DVI connection can never carry audio. Some also call this "digital video interface."
Flat-panel TV	Any ultrathin, relatively lightweight TV—especially those that can be wall-mounted. Current flat-panel TVs use either plasma or LCD screen technology.
Frame	In moving picture media, whether film or video, a frame is a complete, individual picture.
Frame rate	The rate at which frames are displayed. The frame rate for movies on film is 24 frames per second (24 fps). Standard NTSC video has a frame rate of 30 fps (actually 60 fields per second). The frame rate of a progressive-scan video format is twice that of an interlaced scan format. For example, interlaced formats like 480i and 1080i deliver 30 complete frames per second; progressive formats like 480p, 720p, and 1080p provide 60.
Front-projection TV	See Projector.
HDMI (high-definition multimedia interface)	Similar to DVI (but using much smaller connectors), the multipin HDMI interface transfers uncompressed digital video with HDCP copy protection and multichannel audio. Using an adapter, HDMI is backward-compatible with most current DVI connections, although any DVI-HDMI connection will pass video only, not audio.
LCD (liquid crystal display)	Liquid crystal display technology is one of the methods used to create flat-panel TVs. Light is not created by the liquid crystals; a "backlight" behind the panel shines light through the display. The display consists of two polarizing transparent panels and a liquid crystal solution packed in between. An electric current passed through the liquid causes the crystals to align so that light cannot pass through them. Each crystal acts like a shutter, either allowing light to pass through or blocking the light. The pattern of transparent and dark crystals forms the image.
LCoS (liquid crystal on silicon)	A projection TV technology based on LCD. With LCoS, light is reflected from a mirror behind the LCD panel rather than passing through the panel. The control circuitry that switches the pixels on and off is embedded further down in the chip so it does not block the light, which improves brightness and contrast. This multilayered microdisplay design can be used in rear-projection TVs and projectors.

Table 5.1 (Continued)   Glossary of Key Display-Related Terms

Lumen	The unit of measure for light output of a projector. Different manufacturers may rate their projectors' light output differently. "Peak lumens" is measured by illuminating an area of about 10 percent of the screen size in the center of the display. This measurement ignores the reduction in brightness at the sides and corners of the screen. The more conservative "ANSI lumens" (American National Standards Institute) specification is made by dividing the screen into nine blocks, taking a reading in the center of each, and averaging the readings. This number is usually 20–25 percent lower than the peak lumen measurement.
Luminance	The brightness or black-and-white component of a color video signal; determines the level of picture detail.
OLED (organic light-emitting diode)	OLED is an up-and-coming display technology that can be used to create flat-panel TVs. An OLED panel employs a series of organic thin films placed between two transparent electrodes. An electric current causes these films to produce a bright light. A thin-film transistor layer contains the circuitry to turn each individual pixel on and off to form an image. The organic process is called electrophosphorescence, which means the display is self-illuminating, requiring no backlight. OLED panels are thinner and lighter than current plasma or LCD HDTVs and have lower power consumption. Only small OLED screens are available at this time, but larger screens should be available by 2011.
Plasma	Plasma technology is one of the methods used to create flat-panel TVs. The display consists of two transparent glass panels with a thin layer of pixels sandwiched in between. Each pixel is composed of three gas-filled cells or subpixels (one each for red, green, and blue). A grid of tiny electrodes applies an electric current to the individual cells, causing the gas to ionize. This ionized gas (plasma) emits high-frequency UV rays that stimulate the cells' phosphors, causing them to glow, which creates the TV image.
Projector	A video display device that projects a large image onto a physically separate screen. The projector is typically placed on a table or is ceiling-mounted. Projectors, sometimes referred to as front-projection systems, can display images up to 10 feet across, or larger. Old-fashioned large, expensive CRT-based projectors have been replaced by compact, lightweight, lower-cost digital projectors using DLP, LCD, or LCoS technology.
Rear-projection TV	Typically referred to as "big-screen" TVs, these large-cabinet TVs generally have built-in screens measuring at least 40 inches. Unlike the bulky CRT-based rear-projection TVs from years ago, today's "tabletop" rear-projection TVs are relatively slender and light. These TVs use digital microdisplay technologies like DLP, LCD, and LCoS.
Viewing angle	Measures a video display's maximum usable viewing range from the center of the screen, with 180 degrees being the theoretical maximum. Most often, the horizontal (side-to-side) viewing angle is listed, but sometimes both horizontal and vertical viewing angles are provided. For most home theater setups, horizontal viewing angle is more critical.
Widescreen	When used to describe a TV, widescreen generally refers to an aspect ratio of 16:9, which is the optimum ratio for viewing anamorphic DVDs and HDTV broadcasts.

chip" projectors dedicate a chip to each main color: red, green, and blue. While three-chip models are more expensive, they completely eliminate the rainbow effect, which is an issue for some viewers. All projectors using three-imaging devices must have all color-generating devices aligned perfectly so that the red, green, and blue information for each pixel converges on the screen; the single-chip DLP design has an advantage in this context over the three-device systems.

- LCD is a technology that is used in flat-panel TVs.* Light is not created by the liquid crystals; a "backlight" behind the panel shines light through the display. The display consists of two polarizing transparent panels and a liquid crystal solution sandwiched in between. An electric current passed through the liquid causes the crystals to align so that light cannot pass through them. Each crystal acts like a shutter, either allowing light to pass through or blocking the light. The pattern of transparent and dark crystals forms the image. So-called 3LCD systems have three separate LCD panels; red, green, and blue. As light passes through the LCD panels, they can be either open or closed to allow light to pass through or be filtered out; this process results in the projection of the image onto the screen. LCD projectors have reasonably accurate colors because of the use of the separate LCD panels, a sharp image, and good efficiency. Disadvantages of LCD technology include pixilation and the fact that it does not produce absolute black, resulting in less contrast than one would achieve with DLP. An advantage is that high-speed response systems eliminate crosstalk (double-contour ghost images).

- LCoS (liquid crystal on silicon) is a projection technology based on LCD. With LCoS, light is reflected from a mirror behind the LCD panel rather than passing through the panel. The control circuitry that switches the pixels on and off is embedded further down in the chip so it does not block the light, which improves brightness and contrast. This multilayered microdisplay design can be used in rear-projection TVs

---

* LCDs accounted for 70% of the global TV market in 2009, with CRTs being about 23% and PDP being 7% [KIM201001].

and projectors. Vendors use different names for their LCoS technologies—Sony uses SXRD, while JVC uses D-ILA and HD-ILA. The primary driver for the success of LCD products in the home theater market is their price advantage. Also, 3LCD tend to have better color brightness compared to the single-chip DLP products.

5.2.1 Commercial Displays Based on Projection

Projection-based, commercially available video/TV systems reconstruct a 3D image using either a number of projectors; these systems use either front- or rear-projection televisions (FPTVs or RPTVs). Multiple projectors (two or more) are called stacked projectors. At the fundamental level, rear-projection home systems use 3LCD, DLP, or LCoS light-origination elements; single projectors tend to use DLP. The dual stack/two projector system (which can be used with any of the light technologies) has a clear advantage over a single projector: It projects more light as a system. Brightness is one of the crucial elements when projecting 3D since more than 80 percent of light output gets lost when projecting 3D; however, it also needs to have alignment functions. Also, the cost is almost linear in the number of elements (projectors) in the stack. Front projection (FP) is clearly the standard in movie theaters; they are also used in very high-end home systems, but the large majority of home projection systems are rear projection (RP) systems. There are a number of variants of the single/dual (stacked) projection systems.

Newer dual-stack, front-projector systems for professional applications (presentations, corporate conference rooms, education lecture halls, public display environments, and so on) can use built-in Infitec narrowband spectral filters, thus requiring passive (not active) glasses. The Infitec spectral filters feature narrowband red-green-blue (RGB) light-passing characteristics; the company offers two sets that can pass the left- and right-eye images with little cross-talk. (Dolby commercialized this technology in theatrical applications and JVC has used it for its D-ILA projectors.) Portable 3D systems for professional applications have been available for some time, but these require DLP technology and active shutter glasses. If solution providers or end users want to wear passive glasses, they need to move to a dual-stack solution. Most applications, however, place polarizing filters over the projector lenses, but this requires the use of a polarization-preserving (silver) screen. These are heavy and costly, and typically not very portable [VVT200901].

Note: Glasses that use circular polarizing filters have an advantage over linear polarized filters: The viewer can tilt the head without causing an alteration of what each eye can see. With linear polarization, one has a clean picture only if the linear polarized filters are properly aligned (if they are straight and level); moving the filters can cause ghosting. On the other hand, circular polarization works even if the filters are not aligned horizontally.

Dual-stack front-projector systems for professional applications can use off-the-shelf 3LCD projectors (see Figure 5.3 and Figure 5.4). One press-time example used two Epson 1730W projectors that offer WXGA resolution and 3K lumens when unmodified; the projectors are fitted into a chassis to hold the two projectors in a firm and fixed position, with some mechanical adjustments to align the two images. A minor modification to the projectors allows the integrator to place the Infitec filter inside so there are no external filters to manage; these filters reduce the light output to about 20 percent of the unfiltered value, or around 600 lumens per projector. The cost was

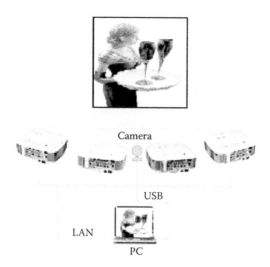

Figure 5.3 (Please see color insert following page 160) Stacked (front) projection concept.

Figure 5.4 Illustrative example of stacked (front) projection. Low-cost solution (Valley View Tech, San Jose, CA, with permission).

around $14,000. According to vendors [ANA200901], "The significant increase in availability of 3D movies . . . is driving demand for dual-stack projector solutions."

Single-projector 3D systems are simpler, less expensive, and more intuitive to use. However, the light output is an ever-present consideration. For home theater systems, the projector's light output is already relatively low, even for 2D systems. Utilizing the same kind of technology will result in much less light output—as noted, typically only 20 percent of the light is effectively usable. Hence, a 3D home theater projector will need to generate a lot more raw light prior to the filtering process—filtering both at the projector and at the viewer's glasses—to achieve the same effective perceived light as a 2D system.

5.2.2 Commercial Displays Based on LCD and PDP Technologies

The previous section discussed projections systems, particularly FP systems that use a projector and a screen. Most consumers prefer an integrated solution. Hence, home theater TVs are typically of the RPTV or flat-panel type. As we noted, the large majority of home projection systems are rear-projection systems. Samsung and Mitsubishi had reportedly shipped over two million RPTVs at press time that can be used to deliver 3DTV content using shutter glasses.

LCD flat-panel TVs and plasma display panel (PDP) TVs are now prevalent in many homes (see Figure 5.5 for a kind of comparison). We described the LCD technology above. A PDP consists of two

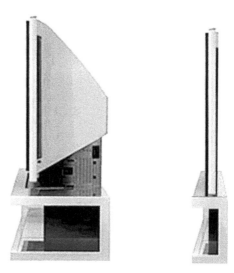

Figure 5.5 Profile view of RPTV and flat panel TV.

transparent glass panels with a thin layer of pixels sandwiched in between. Each pixel is composed of three gas-filled cells or subpixels (one each for red, green, and blue). A grid of tiny electrodes applies an electric current to the individual cells, causing the gas to ionize. This ionized gas (plasma) emits high-frequency UV rays that stimulate the cells' phosphors, causing them to glow, which creates the TV image [KIN200901].* Both of these panel displays have a screen refresh rate. HDTV content typically has a frame rate of 60 Hz and existing displays have been built to support that rate. The stereoscopic 3DTV signal operates at 120 Hz: a frame, 60 times a second, for each eye, or a total frame rate of 120 Hz. Newer systems can now *operate at* 120 or 240 Hz and *accept incoming signals at that rate.*†

---

* Industry observers believe that OLED displays will not be completely viable for commercial HDTV or 3DTV until 2015.

† Some existing 240 Hz screen operate as follows: The content input rate is 60 Hz; therefore the TV's electronics create three interpolated frames for every incoming frame and sequentially display those four frames. For 3DTV (active shutter glasses) to operate properly, that process has to change (has to be eliminated). Here is how the operation must take place: The left-eye incoming frame (arriving at 60 Hz) needs to be displayed for the first subframe and then held during the second subframe while the shutter lens of the left eye of the glasses opens up to allow the left eye to see it. In the third subframe the right-eye image must be grabbed from the incoming signal and written to the screen; it must then be held during the fourth subframe while the shutter lens of the right eye of the glasses opens up to allow the right eye to see it.

LCD-based systems can both be used with temporal interlacing, which require shutter glasses or line interlacing, and which can then use polarized glasses.

For an illustrative example, Panasonic Corporation announced at press time that it had developed a 50-inch, full-HD, 3D-compatible PDP and high-precision active shutter glasses that enable the viewing of theater-quality 3D images in the living room. Aiming to bring full-HD 3DTVs to the market in 2010, the company was stepping up its efforts in developing the related technology. Figure 5.6 depicts the end-to-end system as envisioned by Panasonic.

5.2.3 LCD 3DTV Polarized Display

While the majority of LCD-based displays were initially of the time-multiplex type, requiring active glasses, interlaced LCD systems that only require polarized passive glasses were appearing at press time. These systems offer a flicker-free visual experience and do not require expensive glasses. For example, for illustrative purposes of the technology, note that in 2009 JVC launched a 46-inch, full-HD, 3D LCD monitor of this type—initially for professional use. The flat-screen monitor uses the Xpol polarizing filter method. The Xpol method allocates images for the right and left eyes to the odd- and even-numbered horizontal lines of the screen. When viewed through a pair of dedicated circular polarization glasses, the images displayed on the odd-numbered lines are visible to the right eye but invisible to the left, and vice versa for the even-numbered lines. JVC claims that there is no flicker, which can be visible in systems using LCD shutter glasses, since both right and left images are displayed on the screen at the same time. Also, since the system relies on polarizing lenses, there is no need to power the glasses. Besides the line-by-line method, the monitor also supports the side-by-side method, which arranges the left and right images on both sides of the screen. The monitor's three HDMI input terminals are compatible with standard HD video signals, including 1080/24p, 50p, 60p, 50i, and 60i. Input signals in line-by-line, or side-by-side, format can be displayed as 3D images with the 50i and 60i signals limited to the side-by-side method only. JVC is planning to manufacture 2,000 units a year worldwide. Although they will initially be targeted at those involved in the production and

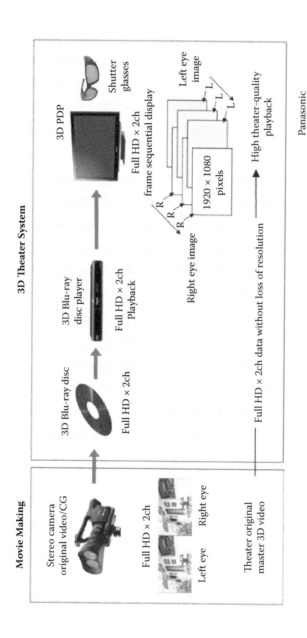

Figure 5.6   End-to-end 3D system as envisioned by Panasonic.

promotion of 3D movies, JVC also sees future applications for the 3D monitor in fields that rely heavily on simulation, such as science, medicine, and education [QUI200901].

Hyundai IT also manufactures a 46-inch LCD, full-HD, polarized 3D display. Additionally LG Electronics has introduced a 47-inch LCD HDTV model that allows viewers to use passive polarized glasses to view the stereoscopic images displayed by the TV. LG's 3D HDTV, available on the consumer market in South Korea since 2009, uses a high-contrast and high-brightness LCD panel that displays high-quality 1080p "full-HD" 2D images and also 3D stereo images that can be delivered over HDMI interface ports in one of many possible formats.

5.2.4 Summary of 3DTV Displays

We summarize here the line of display options available at this time. See Table 5.2, where six distinct approaches are listed.

5.2.5 Glasses Accessories

As discussed, glasses can be passive or active. Active shutter glasses are relatively expensive (in the $50 to $500 range), which can be an issue for a large family or gathering, and also are less efficient at the optical light-intensity level. Polarization-based glasses require a polarization-preserving screen; some of these screens can exhibit optical aberrations.

Active shutter glasses make use of a synchronization (sync) signal from the display set to open and close the right/left LCD in the glasses worn by the viewer in temporal alignment with the screen painting of the left-eye image frames and the right-eye image frames. The link is usually based on infrared transmission, but other approaches have also emerged. For example, Texas Instruments has developed a visible light system that makes use of white pulses embedded in the picture that is used by the glasses to switch between eyes. No standard protocol has emerged as of yet to handle this device-level interaction, especially in reference to the signaling handshake; various companies use proprietary schemes. Some work was underway at

Table 5.2   Summary of Possible, Commercially Available TV Screen/System Choices for 3D

3D DISPLAY SYSTEM	ADVANTAGES	DISADVANTAGES
Projection based FPTV (polarized display) with passive glasses	Big screen 3D effect similar to cinematic experience Excellent-to-good light intensity Choice of projectors/cost Inexpensive lightweight passive 3D glasses	Needs a silver screen to retain polarization of light Alignment of two projectors stacked on top of each other needed Not totally décor-friendly
Projection based FPTV (unpolarized display) with active glasses	Option of using newer single DLP projectors that support 120hz refresh rate (active based system) No polarization-protecting screen needed	More expensive glasses Need battery-powered LCD shutter glasses.
Projection based RPTV (polarized display) with passive glasses	Integrated unit—easier to add to room décor To present a stereoscopic content, two images are projected superimposed onto the same screen through different polarizing filter (either linear or circular polarizing filters can be used) Viewer wears low-cost eyeglasses that also contain a corresponding pair of different polarizing filters	Some light intensity loss at the display level Not of the flat-panel-TV type; cannot be hung on wall
LCD 3DTV (polarized display) with passive glasses	Simple to use system not requiring projection setup To present a stereoscopic content, two images are projected superimposed onto the same screen through interlacing techniques	Some possible loss of resolution Viewer wears low-cost eyeglasses that also contain a pair of different polarizing filters Some light intensity loss at the display level Relatively expensive
3D Plasma/LCD TV (unpolarized display) with active glasses	Simple to use system not requiring projection setup Flat-screen TV type, elegant décor	Deliver two images to the same screen pixels, but alternates them such that two different images are alternating on the screen Active shutter glasses can be expensive, particularly for a larger viewing group

Continued

Table 5.2 (Continued) Summary of Possible, Commercially Available TV Screen/System Choices for 3D

3D DISPLAY SYSTEM	ADVANTAGES	DISADVANTAGES
		Requires TV sets to be able to accept and display images at 120/240 Hz
		Glasses need power
		Some light intensity loss at the viewer (glasses level)
		Some resolution loss
		Size limited 60–80 inches at this time but larger systems being brought to market
Autostereoscopic screen (lenticular or barrier)	No glasses needed	Fewer suppliers
		Further out
		Some key manufactures have exited the business (for now)
		Content production is more complex
		Displays have a "sweet spot" that requires viewers to be within this viewing zone.

press time by the 3D@Home Consortium and the CEA to develop usable standards.

Glasses that are planned to be used for LCD and PDP TV include the following:

- 120 Hz active shutter glasses (for PDP and LCD)
- Xpol patterned retarder (for LCD)
- Active retarder (for LCD) (emerging technology)

Shutter glasses can be used for both PDP and LCD, as just noted. They require that the image on the screen be fully and properly refreshed at 120 Hz. This process is also called "page flipping." In order to eliminate (or at least reduce) ghosting and cross-talk, good timing synchronization is required. The cost of the glasses will be a consideration for some viewers and some situations.

In the Xpol approach one adds a patterned retarder sheet over the 2D LCD panel by employing a lamination process. This retarder consists of rows of retarders and polarizers that are aligned with the rows of the LCD panel. With this arrangement, a row is polarized (say, 45 degrees) in one state and retarded to a right-hand circularly (RHC) polarized state. The next row is polarized in the orthogonal direction (say, 135 degrees) in one state and retarded to a left-hand circularly (LHC) polarized state. The screen reproduces an incoming image in such a manner that the left-eye image appears on, for example, odd rows and the right-eye image appears on even rows. Finally the viewer wears passive polarized glasses separating the two images. A 60 Hz screen can be utilized. One drawback of this approach is that the vertical resolution perceived by each eye is half of what the incoming signal brings and so half of what it is in the shutter glasses case. Some viewers find this objectionable.

The active retarder approach (possibly to be available by 2012) is a hybrid of the two approaches just described. In the shutter approach, the shutter function is placed in front of the eyes and it allows or blocks view based on the synchronization filter. In the active retarder design, the shutters are placed on the LDC screen, allowing the viewer to use passive glasses.

In summary, active glasses have the following drawbacks:

- Require power
- Relatively expensive
- Generally heavier
- May be TV-manufacturer specific (until standards develop for synchronization)
- May introduce flicker: can be visible in systems using shutter glasses, since both right and left images are displayed on the screen at the same
- Reduce effective light intensity arriving at the eye

However, given the fact that they have active components (electronics), they can be enhanced over time to incorporate other advanced functions and features.

5.3 Other Display Technologies

This section identifies some other systems that were being discussed at press time. They tend to be concepts or prototypes. However, they may be examples of systems that could perhaps become the technology of choice a few years down the road.

5.3.1 *Autostereoscopic Systems with Parallax Support in the Vertical and Horizontal Axes*

Most autostereoscopic systems available for consideration for 3DTV (based on lenticular or parallax barrier approaches) only support a left-right parallax variation. For example, a viewer may want to watch TV while he or she is lying down and not lose the 3D sensation when viewing is done in that position. To achieve this capability, one needs parallax support in both the vertical (V) and horizontal (H) axes. Prototypes systems (still based on lenticular approaches) that allow left-right and up-down parallax generation have been recently demonstrated. One such system was demonstrated by Hitachi in 2008 [OIK200801]. In this demo, autostereoscopic display 3D images can be watched regardless of the viewing direction (that is, from above, below, left, or right). The content for these systems, however, tends to be computer-generated (CG) and/or still images (these with photos taken with integral photography means discussed earlier); there is interest in being able to extend the techniques to integral videography; Figure 5.7 depicts the arrangement used.

A plate-type microlens array, composed of hundreds of 300-micron-diameter convex lenses lined up in columns, is attached on top of a LCD. The array is configured such that there are 20 to 80 or so liquid-crystal pixels under each convex lens. Since the angle for viewing the lens is slightly different for the left and right eyes, the magnified liquid crystal pixels appear different when viewed by the left and right eyes. By outputting pictures with separate orientations in regards to each

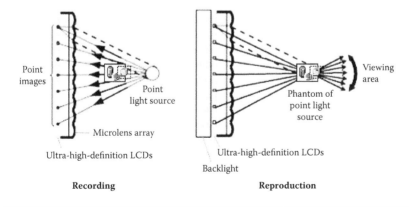

Figure 5.7 (Please see color insert)   Principle of integral photography.

LCD panel   Microlens array

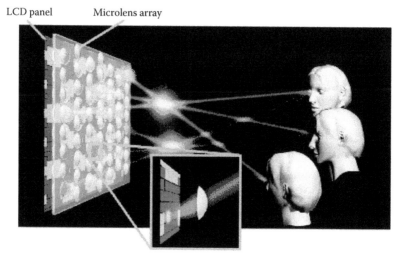

Optimum color-filter arrangement for realizing light-ray-direction
multiplication and moire reduction

Figure 5.8 (Please see color insert)   Display technology to support integral videography.

liquid crystal pixel, this mechanism can show images with different
orientations to the left and right eyes, thereby generating parallax
that in turn creates a stereoscopic sense (see Figure 5.8 from reference
[OIK200801]). In recent years, as a result of advances in manufacturing technology, it has become possible to manufacture fairly good
microlens arrays. Presently, the size of an individual lens has reached
the 300-micron level. Although smaller sizes can be manufactured,
by going too small the pixels of the LCD panel below the microlens array will have to be made smaller too. Getting the right balance
between these sizes is difficult.

The research effort also looked at ways to create contents so that
viewing becomes optimum when they are viewed from an angle of
about 60 degrees in the up and down directions to the image plane.
There are two ways of generating this kind of content. One method
involves shooting images from multiple viewpoints by using multiple
cameras, while the other method entails placing the microlens array
in front of a high-definition camera and shooting through the array.
When shooting with the microlens array in front of the film of a camera, multiple point images are shot, in correspondence with each lens,
through the lens array. The fundamental principle is that when a backlight is shone from the backside of the shot film and looked at through

the microlens array, the multiple point images appear to look like a point light source (this is the principle behind integral photography).

The capture of integral videography material, however, can be very bandwidth intensive because of the need to use ultra-high-definition LCDs. Therefore, the more immediate applications may be in health care—computed tomography (CT) scanning and magnetic resonance imaging (MRI) acquire 3D data and that data can be expressed on the autostereoscopic display.

Apparently as a follow-on from this work, Hitachi announced in late 2009 a full parallax autostereographic 3D TV concept (3D display prototype) it calls a "full parallax 3D TV" using the approach of "integral photography with overlaid projection." (See Figure 5.9 for a pictorial view of the operation.) The 3D display prototype is a 10-inch

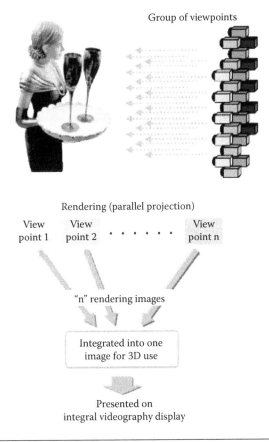

Figure 5.9 (Please see color insert)   Hitachi prototype "integral photography with overlaid projection" concept.

video graphics array (VGA), 640x480 pixel resolution 3D that uses 16 projectors and a lens array sheet to cover them to create the 3D effect. The lens array sheet generates parallax in any direction (not only in the horizontal direction); as a result, the 3D image seen by the user differs in accordance with the angle from which the screen is viewed. The total pixel count of the display is the product of the pixel count in the 3D image by the number of viewpoints showing different images [LCD200901]. The prototype uses multiple small projectors to project overlaid images through the lens sheet on the display panel, and each ray of light radiating from the panel will form the image that should be seen from that particular angle; as a result one can move around the display and see how the object should look like from every angle [MAR200901]. The image resolution can be increased by the number of projectors. Clearly this concept for TV is not ready for commercial content delivery due to the large amount of bandwidth it would use.

5.3.2 Autostereoscopic Systems for PDAs

A 3D technology from 3M that does not require glasses for viewing and that can be used for cell phones, PDAs, laptops, and video games was announced in early 2010 as an OEM product. It uses 3M's enhanced specular reflector (ESR) film, Vikuiti. ESR makes use of 3M multilayer optical film technology to create a highly efficient, specular reflector; it is an ultra-high reflectivity, mirror-like optical enhancement film. Utilizing 3M's multilayer polymer technology, its nonmetallic thin-film construction lends itself to incorporation into a wide variety of configurations and applications. Designed primarily as the rear light-guide reflector to be used with other Vikuiti films for high-efficiency brightness enhancement in light-recycling LCD applications, it can also be used alone in any application requiring a high-performance specular reflector. The film's high reflectivity remains relatively constant across the visible spectrum, producing no unwanted color shifts. Figure 5.10 shows the arrangement of the films, while Figure 5.11 depicts the optical application for 3D displays. 3M believes that the new optical film will enable mass adoption of three-dimensional viewing in handheld devices. The 3M "3D film" integrates into handheld displays and delivers autostereoscopic viewing for extended periods without eye fatigue. Whether there

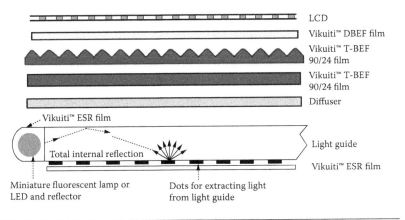

Figure 5.10 ESR film.

will be service providers actually broadcasting 3D content remains to be seen.

The 3D optical film goes into the device's backlight unit and uses two alternate rows of LED lights to sequentially project left and right images into the line of vision. The sequential images are focused on the viewer's eyes. The design is such that the screen brightness and resolution of the original display are not compromised. The surface of the devices features a film with lenses on one side and prism structures on the other that display the corresponding left- and right-eye portions of the image. The panels were initially available in both 9-inch and 2.8-inch formats. Backlight module assembly is nearly identical to existing systems, allowing for simple integration at the assembly stage. The usual optical film stack is replaced with a reflective film,

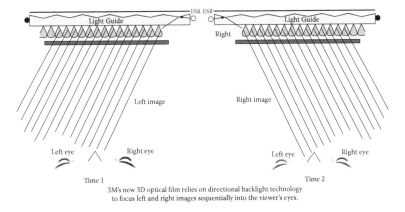

Figure 5.11 Application of ESR to 3D displays.

custom light guide, and 3D film. Through directional backlight technology, left and right images are focused sequentially into the viewer's eyes, allowing full resolution of the display panel. No precision registration during assembly is required.

5.4 Conclusion

Currently there are a number of 3D display systems that can be targeted for the home. The most commercially advanced system (likely to see initial widespread deployment) is the classical two-view stereo display systems requiring the use of glasses. More sophisticated candidates for 3DTV in the home are the multiview autostereoscopic displays, which do not require glasses. These displays emit more than one view simultaneously but the system ensures that viewer only sees a stereopair from a specific viewpoint (available 3D displays are capable of showing nine different images at the same time, of which only one stereopair is visible from a specific viewpoint). Motion parallax viewing can be supported if consecutive views are stereo pairs and are arranged properly [3DP200802].

Appendix 5A: Primer on Cables/Connectivity for High-End Video

This appendix provides an overview of high-end video connectors that may be used in a 3DTV environment.

5A.1 In-Home Connectivity Using Cables

A number of cabling arrangements are in common use.*

5A.1.1 Digital Visual Interface (DVI)   DVI† is a broadly deployed interface technology developed to maximize the quality of flat-panel LCD monitors and modern video graphics cards. It is a replacement for the plug-and-display (P&D) standard, and an improvement from the digital-only DFP format for older flat panels. DVI cables are popular with video card manufacturers, and most cards now manufactured include both a VGA and a DVI output port (see Figure 5A.1). In addition to being used as the standard computer interface, the DVI standard was for a while the digital transfer method of choice for HDTV, plasma display, and other high-end video displays for TV, movies, and DVDs. The digital market is now moving towards the HDMI interface for high-definition media delivery. There are three types of DVI connections: DVI-digital, DVI-analog, and DVI-integrated (digital and analog) [VAN201001]:

- **DVI-D—True Digital Video.** DVI-D cables are used for direct digital connections between source video (namely, video cards) and digital LCD monitors. This provides a faster, higher-quality image than with analog due to the nature of the digital format. All video cards initially produce a digital video signal that is converted into analog at the VGA output. The analog signal travels to the monitor and is reconverted back into a digital signal. DVI-D eliminates the analog conversion process and improves the connection between source and display.

---

* Material in this section is from references [VAN201001], [VAN201002], [VAN201003]. Used by permission by DataPro International Inc.
† Some also use the form "Digital Video Interface."

Figure 5A.1 Digital visual interface (DVI) cable. (@2010 DataPro International Inc. Used by permission.)

- **DVI-A—High-Res Analog.** DVI-A cables are used to carry a DVI signal to an analog display, such as a CRT monitor or budget LCD. The most common use of DVI-A is connecting to a VGA device, since DVI-A and VGA carry the same signal. There is some quality loss involved in the digital to analog conversion, which is why a digital signal is recommended whenever possible.

- **DVI-I—Integration.** DVI-I cables are integrated cables that are capable of transmitting either a digital-to-digital signal or an analog-to-analog signal. This makes it a more versatile cable, being usable in either digital or analog situations.

5A.1.2 High-Definition Multimedia Interface® (HDMI®) HDMI is the current all-in-one, standardized, universal connector for audio/video applications, aiming at unifying all digital media components with a single cable, remote, and interface. According to market research firm In-Stat, over 394 million HDMI-enabled devices were expected to ship in 2009, with an installed base of one billion devices. By the end of 2009, 100 percent of digital televisions are expected to have at least one HDMI input [COL200901]. HDMI products started shipping in 2003. The original HDMI specification had a 5 Gbps bandwidth limit, over twice that of HDTV (that runs at 2.2 Gbps) and is built with the idea of being forwards-compatible by offering unallocated pipeline for future technologies. The maximum pixel clock rate for HDMI 1.0 was 165 MHz, which was sufficient for supporting 1080p and WUXGA (1920x1200) at 60 Hz. The connectors are sliding contact (similar to FireWire and USB) instead of

DataPro

Figure 5A.2 High-definition multimedia interface cable. (@2010 DataPro International Inc. Used by permission.)

screw-on (as is the case in DVI), and are not nearly as bulky as most current video interfaces (see Figure 5A.2). The high bandwidth of HDMI is structured around delivering the best quality digital video *and* audio throughout an entertainment center. The HDMI cable is designed to replace all analog signals (such as S-video, component, composite, and coaxial) as well as HDTV digital signals (such as DVI, P&D, and DFP) with no compromise in fidelity from the source. Additionally, HDMI is capable of carrying up to eight channels of digital-audio, replacing the old analog connections (RCA, 3.5 mm) as well as optical formats (SPDIF, Toslink). HDMI is permeating the home theater market and is on most HDTVs, DVD players, and receivers available today. It has become the standard for entertainment solutions but has made relatively little progress in the computer industry (still dominated by DVI).

HDMI 1.3 increases the bandwidth limit to 10.2 Gbps to allow for the video and audio improvements of the upgraded standard. HDMI 1.3 offers a higher video throughput, at 340 MHz, to allow for higher-resolution displays (such as WQXGA, 2560 × 1600), deep color (up to 48-bit RGB or YCbCr color depths), and the new Dolby standards for loss-less compressed high-definition audio. The HDMI 1.3 specification also designates the use of a smaller connector than the original HDMI connector, dubbed "Mini HDMI." They are both similar in appearance, but the HDMI Mini plug measures about half the size. HDMI 1.3 is already available in commercial products, having started with the release of the Sony PlayStation 3 and various BD players. Models of DVD players, high-definition displays, and A/V receivers released since 2008 are also being designed with HDMI 1.3 connectors [VAN201002].

In mid 2009 an HDMI 1.4 specification was announced. The specification will, among other advancements, offer networking capabilities with Ethernet connectivity and will add an audio return channel to enable upstream audio connections via the HDMI cable. It also includes a 3D capability. The HDMI 1.4 specification will offer the following enhanced functionalities [COL200901], [HDM200901]:

- **HDMI Ethernet Channel.** The HDMI 1.4 specification adds a data channel to the HDMI cable and will enable high-speed bidirectional communication. Connected devices that include this feature will be able to send and receive data via 100 Mbps Ethernet, making them instantly ready for any IP-based application. The HDMI Ethernet channel will allow an Internet-enabled HDMI device to share its Internet connection with other HDMI devices without the need for a separate Ethernet cable. The new feature will also provide the connection platform to allow HDMI-enabled devices to share content between devices (see Figure 5A.3). The HDMI Ethernet channel enables a number of new possibilities via the HDMI link, including:
  - **Sharing an internet connection**—The HDMI Ethernet channel feature allows Internet-ready entertainment devices, from gaming consoles to BD players and more, to share an Internet connection without any need for a separate Ethernet cable.
  - **Content distribution**—Devices connected by the HDMI Ethernet channel will be able to exchange digital content in its native format, enabling recording, storage, and playback options across a connected system, with no need for a separate Ethernet cable.
  - **Home entertainment networking**—The HDMI Ethernet channel accommodates current and future IP-based networking solutions for consumer electronics, such as UPnP, Liquid HD, and DLNA. HDMI with Ethernet is the ideal one-cable solution for connecting devices in these advanced home-networking environments.

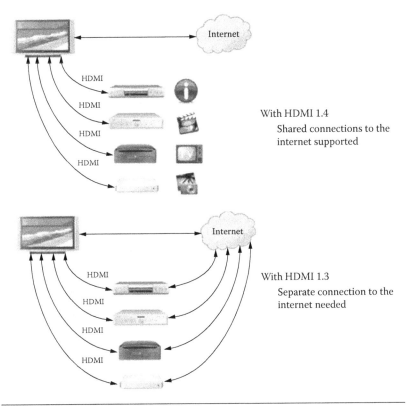

With HDMI 1.4
  Shared connections to the
  internet supported

With HDMI 1.3
  Separate connection to the
  internet needed

Figure 5A.3   HDMI Ethernet channel.

- **Audio Return Channel.** The new specification adds an audio return channel that will reduce the number of cables required to deliver audio upstream for processing and playback. In cases where HDTVs are directly receiving audio and video (A/V) content, this new audio return channel allows the HDTV to send the audio stream to the A/V receiver over the HDMI cable, eliminating the need for an extra cable.
- **3D over HDMI.** The 1.4 version of the specification defines common 3D formats and resolutions for HDMI-enabled devices. The specification will standardize the input/output portion of the home 3D system and will specify up to dual-stream 1080p resolution. The HDMI 1.4 specification establishes protocols for a number of popular 3D display methods, including:
  - Frame, line, or field alternative methods
  - Side-by-side methods (full and half)
  - 2D plus depth methods

- The HDMI 1.4 specification includes information on a wide range of 3D display formats at up to 1080p resolution, including:
  - Field alternative
  - Frame alternative
  - Line alternative
  - Side-by-side half
  - Side-by-side full
  - L + depth
  - L + depth + graphics + graphics depth
- **4K x 2K Resolution Support.** The new specification enables HDMI devices to support HD resolutions four times beyond the resolution of 1080p. Support for 4K x 2K will allow the HDMI interface to transmit content at the same resolution as many digital theaters. Formats supported include:
  - 3840x2160 24Hz/25Hz/30Hz
  - 4096x2160 24Hz
- **Expanded Support for Color Spaces.** HDMI technology now supports color spaces designed specifically for digital still cameras. By supporting sYCC601, Adobe RGB, and AdobeYCC601, HDMI-enabled display devices will be capable of reproducing more accurate lifelike colors when connected to a digital still camera.
- **Micro HDMI Connector.** The micro HDMI connector is a significantly smaller 19-pin connector that supports up to 1080p resolutions for portable devices. This new connector is approximately 50 percent smaller than the size of the existing HDMI mini connector.
- **Automotive Connection System.** The automotive connection system is a cabling specification designed to be used as the basis for in-vehicle HD content distribution. The HDMI 1.4 specification will provide a solution designed to meet the rigors and environmental issues commonly found in automobiles, such as heat, vibration, and noise. Using the automotive connection system, automobile manufactures will now have a viable solution for distributing HD content within the car.
- **Consumers will have a choice** of the following HDMI cables:
  - Standard HDMI cable—supports data rates up to 1080i/60

- High-speed HDMI cable—supports data rates beyond 1080p, including Deep Color and all 3D formats of the new 1.4 specification
- Standard HDMI cable with Ethernet—includes Ethernet connectivity
- High-speed HDMI cable with Ethernet—includes Ethernet connectivity
- Automotive HDMI cable—allows the connection of external HDMI-enabled devices to an in-vehicle HDMI device

5A.1.3 DisplayPort   DisplayPort is the newest digital video interface connector and is designed as a replacement for the DVI standard currently in use on computer hardware; it is similar in specification to the HDMI standard, but unlike HDMI, DisplayPort is being targeted as a computer interface more than a home-theater interface. While VGA has served the computer industry well for many years, its analog-based signals (frequency-modulated red, green, and blue components) are a weakness in fidelity. The digitally rendered signal must be converted to analog by the video card, which adds inconsistencies and blending, and identical source images can vary greatly on different displays, depending on the displays' subjective calibrations. DVI and DisplayPort are both digital-only signals; this ensures that the final image displayed is identical to the image rendered by the computing hardware. Excluding DVI's existing dominance in the market, DisplayPort has a number of advantages over its predecessor [VAN201003]:

- The DisplayPort connector is small and screwless, for easier installation, and added usability in space-conscious hardware
- DVI offers no audio support; DisplayPort offers full digital audio support (up to eight channels) in the same cable as video
- DVI is crippled by its maximum spec length of 5 m, while DisplayPort is designed for up to 15 m.

Unfortunately, DisplayPort and DVI use fundamentally different signal-processing methods, so adapting between the two cannot be done with a simple adaptor or cable (in most circumstances). DVI, like VGA and HDMI, uses separate data channels for each color, requiring a high bandwidth at all times. DisplayPort renders the entire image and breaks it into packets, and these are transferred to the display much like network data over an Ethernet line. Some DisplayPort ports are built to be compatible with DVI internally and can be adapted passively, but this is not a requirement of the DisplayPort standard. In these situations it will appear as though the DisplayPort is being "converted" to DVI, but it is actually the hardware outputting a DVI signal through a DisplayPort connector. If the hardware in use is not capable of outputting this DVI signal, then a DisplayPort-to-DVI adaptor will not function properly.

While HDMI is the digital standard targeted toward home theaters and DisplayPort targeted toward computer electronics, DisplayPort has a feature list virtually identical to HDMI. In fact, the DisplayPort 1.1 standard was adjusted to specifically include the HDCP standard to improve compatibility with HDMI. Because DisplayPort has a newer release schedule than HDMI, it does outperform in a handful of specifications:

- It has a maximum bandwidth of 10.8 Gbps, compared to HDMI at 10.2 Gbps
- It supports the DPCP (DisplayPort content protection) standard in addition to HDCP
- It is an open standard, available to all manufacturers at no cost; HDMI is licensed by HDMI LLC, which raises the cost to manufacturers and consumers.

See Table 5A.1 for some interworking observations.

5A.2 In-Home Connectivity Using Wireless Technology

A number of wireless approaches to connectivity have emerged or are emerging.

5A.2.1 Wireless Gigabit Alliance In late 2009 the Wireless Gigabit Alliance (WiGig) published a specification for 7 Gbps communications over wireless links; WiGig states that products would appear by the end of 2010. The alliance envisions a global wireless ecosystem of interoperable, high-performance devices that work together seamlessly; their proposed technology enables multi-gigabit-speed wireless communications among these devices and aims at industry

Table 5A.1   Interworking between Cable Systems

VGA and DisplayPort Interworking	VGA supports only analog-based signals, and DisplayPort only digital-based. A conversion from VGA to DisplayPort (or vice versa) will require an electronic converter, much like today's VGA to DVI-D units.
DVI and DisplayPort Interworking	Aside from DVI's lack of audio support, the two interfaces are somewhat compatible. DisplayPort uses a signal technology entirely different from DVI/HDMI and is not natively compatible. However, some DisplayPort hardware has built-in convertors to a DVI-compatible format, so a passive DVI/HDMI adaptor will function properly. If the DisplayPort does not have this feature, then external electronic conversion will be necessary.
HDMI and DisplayPort Interworking	DVI and HDMI use the same signal technology, so HDMI suffers the same incompatibility with DisplayPort, and likewise the same possibility for conversion. If the DisplayPort hardware has built-in adaption, then a DisplayPort-to-HDMI cable or adapter will function perfectly. If not, then an external HDMI-DisplayPort convertor will be necessary.

convergence to a single radio using the readily available, unlicensed 60 GHz spectrum [WGA201001].

Proponents make the pitch:

> Make no mistake, Wi-Fi is a terrific system for hooking up multiple computers to each other and to the internet, cable-free. Everybody loves it. But Wi-Fi just won't cut it in the brave new entertainment world of high-definition (HD) movies, delivered on demand via the Internet. . . To get it into your brand-new big-screen Full HD TV, preferably without the need for extra cables snaking across the living room floor, one will need something much, much faster than Wi-Fi. [FRI200901]

The proposal makes use of Wi-Fi's higher-layer protocols, implying that the new technology can be introduced without a substantial change to existing products. The unified wireless specification allows Wi-Fi connections to operate at speeds of 7 Gbps, even while supporting backwards compatibility with existing Wi-Fi devices. That speed is at least 10 times faster than Wi-Fi 802.11n, the speediest version of Wi-Fi, and hundreds of times faster than older Wi-Fi versions. This development will enable the widespread use of the 60 GHz wireless technology and will supplement the 802.11 medium access control (MAC) layer. The 7 Gbps WiGig standard would work at distances of 10 m, which makes it practical for moving high-definition video and data around in open spaces such as living rooms. The downside to such a high frequency is that it transmits reliably only over short distances, perhaps 5 to 10 m. That means WiGig will mainly be used in a single room, rather than broadcasting through a house like Wi-Fi [WGA201002], [FRI200901].

The unified wireless specification enables high-performance wireless display and audio and provides data transfer rates much faster than today's wireless LANs, extending Wi-Fi technology while supporting backward compatibility with existing Wi-Fi devices. The specification paves the way for the introduction of the next generation of high-performance wireless products, including PCs, mobile handsets, TVs and displays, BD players, and digital cameras. The physical layer enables both the low power and the high performance of WiGig devices, which guarantees interoperability and communication at gigabit rates.

As of early 2010, WiGig member companies were planning to begin development work on gadgets that make good use of the new standard. Press-time expectations were that there would be WiGig-enabled PCs, TVs, monitors, videocams, handhelds, BD players, games consoles, and even mobile phones by the end of 2010 [FRI200901]. WiGig has the backing of 30 top technology companies, including chipmakers Intel, AMD, and Nvidia; TV and PC makers such as LG, Samsung, Panasonic, NEC, Dell, and Toshiba; and mobile phone market leader Nokia (Sony was not on the list at press time).

5A.2.2 WirelessHD There are several other systems, including WirelessHD, that use 60 GHz transmissions to flash data between BD players and HDTVs at up to 4 Gbps. This is mainly seen as a cable replacement for BD players. The big push for WirelessHD is coming from Sony, also one of the major forces behind Blu-ray.

The WirelessHD Consortium, led by several leading technology and consumer electronics companies, serves to organize an industry-led standardization effort to define a next-generation wireless digital interface specification for consumer electronics and PC products.* Specifically, the WirelessHD specification will enable wireless connectivity for streaming HD audio, video, and data between source devices and high-definition displays. The WirelessHD specification

---

* This material is based on documentation from the WirelessHD Consortium.

will serve as the first and only wireless digital interface to combine uncompressed high-definition video, multichannel audio, intelligent format and control data, and Hollywood-approved standard content-protection techniques. For end users, elimination of cables for audio and video dramatically simplifies home theater system installation and eliminates the traditional need to locate source devices in the proximity of the display. Also, the technology will support the development of adapter solutions that will be capable of supporting legacy systems.

The WirelessHD specification has been built and optimized for wireless display connectivity, achieving in its first-generation implementation high-speed rates up to 4 Gbps for the consumer electronic, personal computing, and portable device segments. Its core technology promotes theoretical data rates as high as 25 Gbps, permitting it to scale to higher resolutions, color depth, and range. Coexisting with other wireless services, the WirelessHD platform is designed to operate cooperatively with existing, wire-line technologies.

The WirelessHD specification defines a wireless video area network (WVAN) for the connection of consumer electronic (CE) audio and video devices. A key attribute of the WirelessHD system is its ability to support the wireless transport of an uncompressed 1080p A/V stream with a high quality of service (QoS) within a room at distances of 10 m. The requirement for high data throughput at distances of 10 m requires a large allocated frequency spectrum. A large amount spectrum is available on an unlicensed basis in many countries in the 60 GHz band. In North America and Japan, a total of 7 GHz is allocated for use, 5 GHz of which is overlapping. The band 57–64 GHz is allocated in North America while 59–66 GHz is allocated in Japan. In addition, Korea and the European Union have also allowed similar allocations. The regulator agencies allow very high effective transmit power (the combination of transmitter power and antenna gain), greater than 10 W of effective isotropic radiated power (EIRP). However, high EIRP and wide allocated bandwidth will allow high throughput connections that are very directional.

The WirelessHD specification defines a novel wireless protocol that enables directional connections that adapt very rapidly to changes in the environment. This is accomplished by dynamically steering the antenna beam at the transmitter while at the same time focusing the

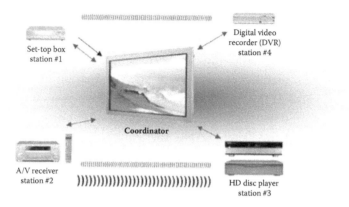

Figure 5A.4 WirelessHD environment.

receiver antenna in the direction of the incoming power from the transmitter. This dynamic beam forming and beam steering utilizes not only the direct path but allows the use of reflections and other indirect paths when the line-of-sight connection is obstructed. This dynamic adjustment of the antenna energy is completed in less than one millisecond.

A WVAN consists of one coordinator and zero or more stations (see Figure 5A.4). The coordinator schedules time in the channel to ensure that the wireless resources are prioritized for the support of A/V streams. The other devices that are a part of the WVAN are referred to as stations. A station may be the source and/or sink of data in the network. The device that is the coordinator also acts as a station in the WVAN and may act as a source and/or sink of data.

Promoters of WirelessHD include Broadcom Corporation, Intel Corporation, LG Electronics Inc., Panasonic Corporation, Royal Philips Electronics N.V., NEC Corporation, Samsung Electronics, Co., Ltd., SiBEAM, Inc., Sony Corporation, and Toshiba Corporation.

5A.2.3 Other Wireless   Wireless Digital Home Interface (WDHI) uses 5 GHz radio spectrum and a video-modem device to stream 1080p high-definition video at a maximum 3 Gbps up to 30 m—enough to cover most houses. The 5 GHz frequency is also used by Wi-Fi 80.11n, which could cause congestion [FRI200901].

References

[3DP200802] 3DPHONE, Project no. FP7–213349, Project title ALL 3D Imaging Phone, 7th Framework Programme, Specific Programme "Cooperation," FP7-ICT-2007.1.5—Networked Media, D5.1—Requirements and specifications for 3D video, 19 August 2008.

[ANA200901] P. Anast, NEC Display Solutions Announces Stunning Stackable Installation Projectors, Press Release, Chicago, 16 June 2009.

[CHI200901] C. Chinnock, 3D coming home in 2010, 3D@Home White Paper, 3D@Home Consortium, www.3Dathome.org

[COL200901] G. Collier and S. Giusti, HDMI licensing, LLC announces features of the upcoming HDMI specification version 1.4—Enhancements include networking, audio return channel, 3D capability, improved performance and new connector, Ogilvy Public Relations for HDMI Licensing, LLC, Sunnyvale, CA, 27 May 2009.

[FRI200901] D. Frith, WiGig puts Wi-Fi systems in shade, *The Australian*, 15 December 2009.

[KIM201001] M. Kim, K. Takanza, 3DTV new battleground for plasma, LCD display makers, Reuters, April 19, 2010.

[KIN200901] S. Kindig, TV and HDTV glossary, Crutchfield, Charlottesville, VA, 2 December 2009.

[LCD200901] Straight from CEATEC 2009: Hitachi's full parallax 3D TV, LCD TV Reviews On Line, 19 October 2009.

[MAR200901] M. Marquit, 3D TV—Without the glasses (w/ video), *PhysOrg. com*, 29 October 2009.

[MAY200901] T. Miyashita, Trends of the flat panel display (FPD) technology which support 3D—Liquid crystal displays (LCD), 3D consortium study session organized by the Technical Sub-committee, Fun Theater in Headquarters Mirai-Kenkyusho of NAMCO BANDAI Games, Higashi-Shinagawa, 22 May 2009.

[HDM200901] HDMI Licensing, LLC, Sunnyvale, CA.

[OIK200801] M. Oikawa, An Interactive Autostereoscopic Display, Hitachi white paper, 11 March 2008.

[ONU200601] L. Onural, A. Smolic, et al., An assessment of 3DTV technologies, *2006 NAB BEC Proceedings*, 456ff.

[POW200901] E. Powell, The Technology War: LCD vs. DLP, ProjectorCentral. com Online Whitepaper, Projector Central, Portola, CA, 28 July 2009.

[QUI200901] D. Quick, JVC launches flicker-free 3D TV, *GizMag Online Magazine*, www.gizmag.com., 6 May 2009.

[SHI200901] A. Shilov, Blu-Ray Disc Association finalizes stereoscopic 3D specification: Blu-Ray 3D spec finalized: New players incoming, *xbitslabs Online Magazine*, 18 December 2009 http://www.xbitlabs.com

[VAN201001] A. van Winkle, *A Complete Guide to the Digital Video Interface*, DataPro International Inc. Seattle, WA.

[VAN201002] A. van Winkle, *What Is HDMI ?* DataPro International Inc. Seattle, WA.

[VAN201003] A. van Winkle, *An Introduction to the DisplayPort Interface*, DataPro International Inc. Seattle, WA.

[VVT200901] Valley View Tech, Valley View Bows Dual-Stack 3D Projectors, 2009, San Jose, CA.

[WGA201001] Wireless Gigabit Alliance (WiGig), http://wirelessgigabitalliance.org

[WGA201002] Wireless Gigabit Alliance (WiGig), New WiFi spec provides speeds up to 7 Gbps, 14 December 2009, http://wirelessgigabitalliance.org

6

3DTV Advocacy and System-Level Research Initiatives

This chapter provides a survey of advocacy activities to support the deployment of 3DTV. As noted in Chapter 1, there currently is a lot of industry interest in this topic. For example, the MPEG of ISO/IEC is working on a coding format for 3D video; as far back as 2003, the 3D Consortium, with 70 partner organizations, was founded in Japan, and more recently, many new activities have been started. Some of these activities are covered in this chapter. In addition to documenting the working groups themselves, the material in this chapter highlights some of the broader technical issues affecting the development of 3DTV services that may or may not have been covered in earlier chapters.

6.1 3D Consortium (3DC)

The 3D Consortium was established in Japan in 2003 by five founding companies (Itochu, NTT Data, Sanyo Electric Company, Sharp, and Sony) and 65 other companies, including hardware manufacturers, software vendors, contents vendors, contents providers, systems integrators, image production, broadcasting agencies, and academic organizations. The 3D Consortium aims at developing 3D stereoscopic display devices and increasing their take-up, promoting expansion of 3D contents, improving distribution, and contributing to the expansion and development of the 3D market.

6.2 3D@Home Consortium

The 3D@Home Consortium was formed in 2008 with the mission to speed the commercialization of 3D into homes worldwide and provide the best possible viewing experience by facilitating the development of standards, road maps, and education for the entire 3D industry—from content and hardware and software providers to consumers [3DA201001].

6.3 3D Media Cluster

3D Media Cluster is the main umbrella structure embracing related European Commission-funded projects to develop joint strategic goals towards 3D Media in the context of future Internet. The work was undertaken in 2004 and in some cases continuing until 2012. Industry consortia and standardization bodies rely on R&D projects that develop the new technologies leading to new standards, products, and markets. Projects under the 3D Media Cluster include 3DTV, 3D4YOU, 2020 3D Media, Mobile3DTV, 3DPhone, 3DPresence, MUTED, and HELIUM3D. Some of these are reviewed below [3DM201001]. All of the material included in this section is based on material from the respective projects/organizations.

Note: While the material that follows in this section looks at activities that originated in Europe, the technical issues are obviously pertinent and applicable at a broad level.

6.4 3DTV

3DTV is a Network of Excellence (NoE) funded by the European Commission Sixth Framework Information Society Technologies Program.* The NoE on 3D television aimed at bringing together a multidisciplinary group of 21 leading European research centers that jointly work on the solution to the challenging technical problems. The consortium, led by Bilkent University, worked for a 48-month period (2004–2008) on 3DTV. The project partners have split their research effort into seven major research work packages. The goal is to align European researchers working in distinct, yet complementary, areas in order to integrate 3DTV research efforts in Europe. All technical building blocks of 3D television, including capture, transmission, and display, are in the technical scope of the project.

---

* This section is based on references [3DT200401], [3DT200601], [3DT200801].

The primary goal of the 3DTV NoE team is to deal with all aspects of the 3DTV in an integrated manner. The team believes that the timing is ideal in terms of technological environment and consumer attitude and needs. The key objective of this project is to align European researchers with diverse experience and activity in distinct, yet complementary, areas so that an effective network for achieving full-scale 3D video capabilities integrated seamlessly to a more general information technology base (like the Internet) is established and kept functional for a long time. The project will create a highly needed synergy among the European partners at a critical time, since 3DTV-related research has been significantly accelerating throughout the world, and therefore will boost European competitiveness. Potential application areas and social impacts of 3DTV will also be investigated.

The things that have prevented such a highly desirable mode of communications from becoming a reality are essentially technological deficiencies. It is not too difficult to identify these missing technological building blocks; for example, the lack of practical, fully electronic means of 3D scene capture and 3D scene display units is probably the main complicating factor. A careful analysis reveals that there are many other technological components missing, and therefore 3DTV is still not a common tool of daily life as in the case of conventional TV. However, the technological components that are necessary to bring 3DTV into reality have significantly matured over the past two decades. For example, image-processing algorithms have evolved to handle video data from multiple synchronous sources and can extract and match feature points from each such source; this paves the road to a successful capture of accurate 3D scene information.

Computer graphics technology has matured to provide almost all necessary tools for abstract 3D scene representation, such as deformable meshes and other generic 3D motion object representation tools. Another main building block is the digital TV technology in broad sense. The past decade has witnessed many important technological jumps in that regard: Major technological advances and standardization activities first gave us videoconferencing and videophone, and then made MPEG-1 (VCD) and MPEG-2 (Digital TV and DVD) a reality, and eventually evolved to the most complicated intellectual property developed in the history, the MPEG-4.

The developed technological base is now ready to be adapted to 3D technology. Telecommunications in general, and Internet protocols in particular, paved the way for easy-to-generate/use video content that can be delivered digitally to our monitors. Streaming video know-how has developed significantly. In the meantime, major technological breakthroughs in optical display technologies have been witnessed: Spatial light modulators (SLM), digital micromirror devices (DMD), acousto-optic technology, and similar approaches have hinted at successful electronic holographic displays. Signal-processing tools are mature enough to tackle all associated fast signal conversion steps needed during the operational phases as signals are captured, processed, and directed to next components of the 3DTV chain up to the display end.

Looking at this confluence collectively, it can be concluded that the scientific and technological environment is ripe for the important enabling step towards 3DTV. This observation is also confirmed by various already established research activities in 3DTV field in Europe (for example, ATTEST; see below), Japan, and the Untied States. An important indicator is the recent activity in MPEG-4 standardization group toward incorporating 3D objects into object-based video technology.

Some results announced along the way by the NoE include the following:

- Capture
    - Many experimental multicamera capture systems were designed and tested. Synchronization among the cameras is achieved.
    - Many techniques were developed to generate automated 3D personalized human avatars from multicamera video input.
    - Image-based methods were developed for surface reconstruction of moving garments from multiple calibrated video cameras.
    - A method based on synthetic aperture radar techniques was developed to increase the resolution of charge-coupled-device (CCD) based holographic recordings.

- Signal-processing methods were developed for automated detection of faces, facial parts, facial features, and facial motion in recorded video.
- A method for generating and animating a 3D model of a human face was developed.
- Representation
  - A method to represent 3D objects using multi-resolution tetrahedral meshes was developed.
  - A technique was developed to recognize head and hand gestures; the method is then used to synthesize speech-synchronized gestures.
  - A method for representing scalable 3D image-based video objects was developed.
  - Software tools for easy description of 3D video objects were developed.
- Coding and Compression
  - A technique to automatically segment stereo 3D video sequences was developed.
  - A full end-to-end multiview video codec was implemented and tested.
  - A storage format for 3D video was developed.
  - A proposal submitted to MPEG for multiview video coding was tested and performed best in subjective tests among eight other proposals from different parts of the world.
  - Multiview test data sets using arrays of eight cameras have been produced and made available to MPEG and general scientific community.
  - Various 3D mesh compression, watermarking, hologram compression techniques, and methods for coding and rendering free-viewpoint video were developed.
- Transmission
  - An optimal cross-layer scheduling for video streaming was developed.
  - An optimal streaming strategy under rate and quality constraints was developed.
  - Different approaches for error concealment in stereoscopic images were developed.

- Color and depth representation for end-to-end 3DTV was further developed and tested.
- Signal Processing Issues in Diffraction and Holography
  - Analytical solutions for complex coherent light-field generation by a deflectable mirror array device were developed.
  - Fast methods to compute diffraction between tilted planes were developed and tested.
  - Algorithms to compute 3D optical fields from data distributed over 3D space were developed and tested.
- Display
  - Autostereoscopic displays for 3DTV were further developed.
  - Viewer tracking autostereoscopic displays were further developed.
  - Characterization and calibration techniques for various spatial light, modulator-based holographic displays were developed.

6.5 Challenges and Players in the 3DTV Universe

The material that follows is included to illustrate some of the issues, concepts, approaches, challenges, opportunities, technologies, and players involved in 3DTV as of press time. The inclusion of this material should not be construed as giving emphasis to the work being developed in any particular geographical region of the world, because, in fact, as noted, there is global R&D work being undertaken at this time on 3DTV.

6.5.1 *European Information Society Technologies (IST) Project "Advanced Three-Dimensional Television System Technologies" (ATTEST)*

This is a project where industries, research centers, and universities have joined forces to design a backwards-compatible, flexible, and modular broadcast 3DTV system.* In contrast to former proposals that often relied on the basic concept of "stereoscopic" video—that is, the capturing, transmission, and display of two separate video streams (one for the left eye and one for the right eye)—this activity focuses

---

* This section is based on reference [FEH200401].

on a data-in-conjunction-with-metadata approach. At the very heart of this new concept is the generation and distribution of a novel data-representation format that consists of monoscopic color video and associated per-pixel depth information. From these data, one or more "virtual" views of a real-world scene can be synthesized in real time at the receiver side (that is, a 3DTV set-top box) by means of depth-image-based rendering (DIBR) techniques. The modular architecture of the proposed system provides important features, such as backwards-compatibility to today's 2D digital TV, scalability in terms of receiver complexity, and adaptability to a wide range of different 2D and 3D displays [FEH200401].

6.5.1.1 3D Content Creation   For the generation of future 3D content, novel three-dimensional material is created by simultaneously capturing video and associated per-pixel depth information with an active range camera such as the Zcam developed by 3DV Systems. Such devices usually integrate a high-speed pulsed infrared light source into a conventional broadcast TV camera and they relate the time of flight of the emitted and reflected light walls to direct measurements of the depth of the scene. However, it seems clear that the need for sufficient high-quality, three-dimensional content can only partially be satisfied with new recordings. It will therefore be necessary (especially in the introductory phase of the new broadcast technology) to also convert already existing 2D video material into 3D using "structure from motion" algorithms. In principle, such methods (either offline or online) process one or more monoscopic color video sequences to (1) establish a dense set of image-point correspondences from which information about the recording camera as well as the 3D structure of the scene can be derived, or (2) infer approximate depth information from the relative movements of automatically tracked image segments. Whatever 3D content generation approach is used in the end, the outcome in all cases consists of regular 2D color video in European digital TV format (720x576 luminance pels, 25 Hz, interlaced) and an accompanying depth-image sequence with the same spatiotemporal resolution. Each of these depth images stores depth information as 8-bit gray values with the gray level 0 specifying the furthest value and the gray level 255 defining the closest value. To translate this data representation format to real, metric depth values (which are required for the "virtual" view generation) and to be flexible with respect to 3D scenes with different depth characteristics, the gray values are normalized to two main depth-clipping planes.

6.5.1.2 3D Video Coding   To provide the future 3DTV viewers with 3D content, the monoscopic color video and the associated per-pixel depth information have to be compressed and transmitted over the conventional 2D digital TV broadcast infrastructure. To ensure the required backwards-compatibility with existing 2DTV set-top boxes, the basic 2D color video has to be encoded using the standard MPEG-2 as MPEG-4 visual or advanced video coding (AVC) tools currently required by the Digital Video Broadcast (DVB) Project in Europe.

6.5.1.3 Transmission   The DVB Project, a consortium of industries and academia responsible for the definition of today's 2D digital TV broadcast infrastructure in Europe, requires the use of the MPEG-2 systems-layer specification for the distribution of audio/visual data via cable (DVB-C), satellite (DVB-S), or terrestrial (DVB-T) transmitters.

6.5.1.4 Virtual-View Generation and 3D Display   At the receiver side of the proposed ATTEST system, the transmitted data is decoded in a 3DTV set-top box to retrieve the decompressed color video- and depth-image sequences (as well as the additional metadata). From this data representation format, a DIBR algorithm generates "virtual" left- and right-eye views for the three-dimensional reproduction of a real-world scene on a stereoscopic or autostereoscopic, single or multiple user, 3DTV display. The backwards-compatible design of the system ensures that viewers who do not want to invest in a full 3DTV set are still able to watch the two-dimensional color video without any degradations in quality using their existing digital 2D TV set-top boxes and displays.

6.5.2 3DPhone

The 3DPhone project aims to develop technologies and core applications enabling a new level of user experience by developing end-to-end, all-3D imaging mobile phones.* Its aim is to have all fundamental functions of the phone—media display, user interface (UI), personal information management (PIM) applications—realized in 3D. This includes mobile stereoscopic video, 3D UIs, 3D capture/content creation, compression, rendering, and 3D display. The research and develop algorithms for 3D audio/visual applications including personal communication, 3D visualization, and content management will be developed. The goal is to enable users to:

- Capture "memories" in 3D and communicate with others in 3D virtual spaces
- Interact with their device and applications in 3D
- Manage their personal media content in 3D

The 3DPhone Project started on February 1, 2008. The duration of the project is three years, and there are six participants from Turkey, Germany, Hungary, Spain, and Finland. The partners are Bilkent University, Fraunhofer, Holografika, TAT, Telefonica, and University of Helsinki. 3DPhone is funded by the European Commission's ICT program in Framework Programme Seven.

The expected outcome will be simpler use and a more personalized look and feel. The project aims at advancing the state of the art in mobile 3D technologies by:

- Implementing a mobile hardware and software platform with both 3D image capture and 3D display capability, featuring both 3D displays and multiple cameras. The project will evaluate different 3D display and capture solutions and will implement the most suitable solution for hardware-software integration.
- Developing user interfaces and applications that will capitalize on the 3D autostereoscopic illusion in the mobile handheld environment. The project will design and implement 3D and zoomable UI metaphors suitable for autostereoscopic displays.
- Investigating and implementing end-to-end 3D video algorithms and 3D data representation formats, targeted for 3D recording, 3D playback, and real-time 3D video communication.

---

* This section is based on reference [3DP200801] and [3DP200901].

- Performing ergonomics and experience testing to measure any possible negative symptoms, such as eyestrain, created by stereoscopic content. The project will research ergonomic conditions specific to the mobile handheld usage, and in particular to the small screen, one hand holding the device, absence of complete keyboard, and limited input modalities.

In support of these activities 3DPhone published two major reports, as follows:

D5.1 Requirements and Specifications for 3D Video. This deliverable first introduces different 3D video application scenarios on mobile phones, such as real-time 3D video communication or 3D video playback. Besides general low complexity requirements that always apply for mobile phone video applications, the different 3D video application scenarios impose distinct requirements. Consequently, different 3D video formats are introduced along with related algorithms, which are available and under development. The different 3D video formats have different advantages and drawbacks. An initial assessment of the 3D video formats versus applications scenarios is given, analyzing which format might be suitable for which application scenario.

D5.2 First Study Results for 3D Video Solutions. This document reports the first study results for 3D video solutions. The different video formats that were defined in D5.1 are used for this, including three depth-based formats, namely single- and multiview video plus depth and layered depth video, and conventional stereo video. Study results and solutions presented in this deliverable address analysis and synthesis as well as coding algorithms for 3D video. Approaches for conventional stereo video are analyzed. For video plus depth (V+D), results on stereo video rendering for mobile phone video applications are evaluated. For multiview video plus depth (MV+D), the report presents an extensive study on depth coding and appropriate evaluation methods as well as high-quality rendering algorithms. Furthermore, extraction and rendering algorithms for layered depth video are investigated. Finally, conclusions and future prospects of study results for 3D video solutions are presented.

In summary, the general requirements on 3D video algorithms on mobile phones are as follows:

- Low power consumption
- Low complexity of algorithms
- Limited memory/storage for both RAM and mass storage
- Low memory bandwidth
- Low video resolution
- Limited data transmission rates, limited bit rates for 3D video signal

These strong restrictions derived from terminal capabilities and from transmission bandwidth limitations usually result in relatively simple video-processing algorithms to run on mobile phone devices. Typically video-coding standards take care of this by specific profiles and levels that only use a restricted and simple set of video-coding algorithms and low-resolution video. The H.264/AVC baseline profile, for instance, only uses a simple subset of the rich video-coding algorithms that the standard provides in general. For 3D video the equivalent of such a low-complexity baseline profile for mobile phone devices still needs to be defined and developed. Obvious requirements of video processing and coding apply for 3D video on mobile phones as well, such as:

- High coding efficiency (taking bit rate and quality into account)
- Requirements specific for 3D video that apply for 3D video algorithms on mobile phones, including
  - Flexibility with regard to different 3D display types
  - Flexibility for individual adjustment of 3D impression

6.5.3 Mobile3DTV

Mobile 3DTV Content Delivery Optimization over DVB-H System (Mobile3DTV) is a three-year project partly funded by the European Union Seventh RTD Framework Programme in the context of the Information and Communication Technology (ICT) Cooperation

Theme and its objective 1.5 Networked Media.* The project started on 1 January 2008. The main objective of Mobile3DTV is to demonstrate the viability of the new technology of mobile 3DTV. The project is developing a technology demonstration system for the creation and coding of 3D video content, its delivery over DVB-H, and display on a mobile device.

The consortium of Mobile3DTV is formed by three universities, a public research institute, and two subject matter experts (SMEs) from Finland, Germany, Turkey, and Bulgaria. The project partners are Tampere University of Technology, Technische Universität Ilmenau, Middle East Technical University, Fraunhofer Heinrich-Hertz-Institute, Multimedia Solutions Ltd., and Tamlink Innovation-Research-Development Ltd. The Mobile3DTV project cooperates closely with a number of public and private organizations, projects, and networks. It is one of the projects within the 3D Media Cluster, an umbrella structure built upon related EC-funded projects to develop joint strategic goals. The consortium of Mobile3DTV also collaborates on project activities with industrial players such as Texas Instruments and DiBcom as well as research institutions such as Nokia Research Center and Electronics and Telecommunications Research Institute of Korea, which are conducting active research in the same area.

The project has been looking at autostereoscopic displays. As discussed in this text, there is a wide range of 3D display technologies, but not all of them are appropriate for mobile use. For example, wearing glasses to aid the 3D perception of a mobile device is highly inconvenient. The limitations of a mobile device, such as screen size, CPU power, and battery life, limit the choice of a suitable 3D display technology. Another important factor is backward compatibility: A mobile 3D display should have the ability to be switched back to 2D mode when 3D content is not available.

As we noted in Chapter 3, autostereoscopic displays present multiple views to the observer, each one seen from a particular viewing angle along the horizontal direction. However, the number of views comes at the expense of resolution and brightness loss—and both are limited on a small-screen, battery-driven mobile device. Because

---

* This section is based on reference [3MO200901].

mobile devices are normally watched by only one observer, two independent views are sufficient for satisfactory 3D perception. At the
moment, there are only a few vendors with announced prototypes of
3D displays targeted for mobile devices. All of them are two-view,
thin-film-transistor (TFT) based autostereoscopic displays.

6.5.4 Real3D

Real3D—digital holography for 3D and 4D real-world objects' capture, processing, and display project—is a research project funded
under the Information and Communication Technologies theme of
the European Commission's Seventh Framework Programme, and
brings together eight participants from academia and industry.* This
three-year project marks the beginning of a long-term effort to facilitate the entry of a new technology (digital holography) into the three-
dimensional capture and display markets. The project duration spans
2008 through 2011.

As we covered in Chapter 3, holography is an inherently 3D technique for the capture of real-world objects and is unrivalled in high-
end industry and scientific research for noncontact inspection of
precision 3D components and microscopic 3D samples. Many existing 3D imaging techniques are based on the explicit combination of
several 2D perspectives (or light stripes and other similar aspects).
The advantage of holograms is that multiple 2D perspectives can be
optically combined in parallel in one step independent of the hologram size. Recently digital holography (DH) (holography using
a digital camera) has become feasible due to advances in scientific
camera technology. The advantage of a digital representation of holograms is that they can be processed, analyzed, and transmitted electronically. Digital holograms, as a new form of digital media, are now
beginning to be processed and analyzed like conventional images. The
participants in this project have pioneered aspects of the capture and
processing of digital holograms. As a generalization of conventional
images, digital holograms have greater potential for wider society than
simply in high-end industry and scientific research. However, in order
for the envisaged applications of holographic image processing (3D

---

* This section is based on reference [REA200901].

display and video, noncontact automated inspection, medical imaging, 3D microscopy, security, 3D computer games, and virtual reality) to become commonplace, it needs to be demonstrated that digital holography can come out of the laboratory and can be usefully and practically employed in day-to-day life.

The principle of digital holography is identical to the classical one. The idea is always to record the interference between an object wave and a reference wave in an in-line or off-axis geometry. The major difference consists in replacing the photographic plate by a digital device like a charged-coupled-device (CCD) camera. Therefore, the wave front is digitized and stored as an array of zeros and ones in a computer and the reconstruction process is achieved numerically through a numerical simulation of wave propagation. This method suppresses the long intermediate step of photographic plate development between the recording and the numerical reconstruction process and allows high acquisition and reconstruction rates. On the other hand, a drawback also exists. In fact, no electronic device is able to compete with the high resolution (up to 5000 lines/mm) of the photographic emulsions used in optical holography. In most cases, the resolution of DH is too low and not qualified for practical applications; however, the computational image reconstruction from a digital hologram has many advantages in respect to the traditional optical holography, enabling processes that are optically not possible. In fact, the amplitude and phase information are obtained separately, allowing amplitude and phase imaging. DH can be useful to perform 3D imaging of the object by retrieving quantitative measurements of some properties as sample thickness, shape profiles or refraction index mapping. Moreover, the use of the DH methods allows other digital wave-front manipulations such as the numerical elimination of the zero order of diffraction or the interferometric comparison of the wave fields generated and reconstructed with different wavelengths. For this reason, DH has become a very useful technique for optical metrology in experimental mechanics, biology, fluid dynamics, and nondestructive inspections. Some limitations, however, arise from the fact that the derivation of the complex wave field from the hologram does not restore perfectly the complex wave field. The positive definite character of the hologram intensity does not provide full bi-univocal correspondence between hologram intensity and complex wave field when the signal is corrupted by parasitic signals or noise either due to the camera or to the optical setup.

We have noted earlier in the text that current and newly developed 3D displays have the disadvantage that they either force the user to wear special eyewear, limit the number of simultaneous viewers, discard completely certain depth cues (such as blurring) and thus causing fatigue, or else encode only a small number of distinct different views of the 3D scene. It can be argued that there is only one known technology that can capture a full 3D scene in a single shot, including phase information, and re-project that light field perfectly thus overcoming all of the above disadvantages: holography. Unfortunately, conventional holograms are not dynamic. But by replacing the conventional holographic plate with a digital camera and an optoelectronic 2D screen, one can capture and display holographic video. However, the full implications of bringing a digital version of holography into the world of 3D video acquisition and 3D display, or how effective it would be, are as yet unknown. The full 3D information encoded in digital holograms has not yet been exploited.

The Real3D initiative aims at eliminating the current obstacles in achieving the world's first fully functional 3D video capture and display paradigm for unrestricted viewing of real-world objects that employs all real 3D principles, hence the acronym "Real3D." Outputs will include building a digital 3D holographic capture, processing, and display arrangement that encompasses the full 360 degrees of perspectives of the 3D scene.

6.5.5 HELIUM3D (High Efficiency Laser Based Multi User Multi Modal 3D Display)

The HELIUM3D project builds upon and advances key technologies developed in the European Union–funded 3DTV Network of Excellence, the MUTED (Multi-User3D Television Display) and the ATTEST (Advanced Three-Dimensional Television Systems) projects.* HELIUM3D seeks to develop a 3D display that will extend the state of the art in autostereoscopic (glasses-free) displays. The HELIUM3D technology addresses the efficiency and color limitations

---

* This section is based on reference [HEL200901].

of current and next-generation displays by developing a new display technology based on direct-view red-green-blue (RGB) laser projection via a low-loss, transparent display screen to the eyes of viewers. The fundamental features of the display are:

- Support for multiple viewers
- Allow for viewer freedom of movement
- Motion parallax to all viewers
- High brightness and color gamut
- Viewer gesture/interaction tracking
- User-centered design, ensuring that future products are "fit for purpose" in terms of perception and usability

Several viewing modes are possible, including motion parallax (the "look around" capability) to each viewer, privacy of viewing from other viewers, a different camera viewpoint to each viewer, and also conventional 2D to all viewers, providing backward compatibility when necessary. This gives a display with a wide range of applications and modes of operation. The display and its viewer-interaction technology will serve both consumer and professional applications. 3DTV and 3D video gaming are the most important consumer applications in which viewer gesture recognition can be a natural replacement for remote controls and game controllers.

HELIUM3D will use pupil tracking to control and direct the light output to the viewers' eyes. The display operates by steering the horizontal direction of light emitted from the surface of the screen where columns of the image are scanned horizontally. High accuracy, low latency pupil tracking, and a high display frame rate permit many modes of autostereoscopic operation to multiple mobile viewers due to the ability of the display to present a different image to every eye.

The display also has a broad potential in professional applications, including medical imaging, video conferencing, engineering design, and oil and gas exploration. In these applications, gesture recognition can be the basis for device-less, hands-free interaction within the 3D object viewing space.

Helium3D will lay the foundations for an autostereoscopic video display capable of being developed into a product that will be in use within the next ten years.

6.5.6 The MultiUser 3D Television Display (MUTED)

The MUTED project aims to develop a practical 3DTV system, which has not been achieved before. The project is supported by funding from the European Commission's Framework Six Programme. The MUTED project aims to produce the first 3D TV display capable of supporting multiple mobile viewers simultaneously and without the need for 3D glasses. No existing 3D display has successfully met all of these requirements that are considered essential for a practical 3D television system. The project will also be investigating ways in which 3D technology can enhance medical scans, allowing doctors and scientists to explore the resulting images in greater detail using 3D displays. Dr. Ian Sexton, leader of the coordinating research group at De Montfort University, notes,

> Three-dimensional televisions have been developed before, but they have all had limitations. This project is a major advance in that we aim to produce a television that is, for the first time, practical. This will be a big step towards people being able to view 3Dimensional television in the comfort of their own homes. It will also explore the potential of the technology to help medical professionals in the diagnosis and treatment of patients by using 3D displays to view MRI and CAT scans, allowing the images to be examined in greater depth. [MUT200901]

There are six other participants in the consortium: Fraunhofer HHI, Germany; the Eindhoven University of Technology, the Netherlands; University of West Bohemia, Czech Republic; Sharp Laboratories of Europe; Biotronics3D; and Light Blue Optics.

6.5.7 3D4YOU

3D4YOU is funded under the ICT (Information and Communication Technologies) Work Programme 2007–2008, a thematic priority for research and development under the specific program "Cooperation" of the Seventh Framework Programme (2007–2013).* The objectives of the project are:

---

* This section is based on reference [3D4200901].

- To deliver an end-to-end system for 3D high-quality media
- To develop practical multiview and depth capture techniques
- To convert captured 3D content into a 3D broadcasting format
- To demonstrate the viability of the format in production and over broadcast chains
- To show reception of 3D content on 3D displays via the delivery chains
- To assess the project results in terms of human factors via perception tests
- To produce guidelines for 3D capturing to aid in the generation of 3D media production rules
- To propose exploitation plans for different 3D applications

The 3D4YOU project aims at developing the key elements of a practical 3D television system, particularly the definition of a 3D delivery format and guidelines for a 3D content creation process. We have mentioned this project and some of its results at various points in this text.

The 3D4YOU project will develop 3D capture techniques, convert captured content for broadcasting, and develop 3D coding for delivery via broadcast, that is, suitable to transmit and make public. 3D broadcasting is seen as the next major step in home entertainment. The cinema and computer games industries have already shown that there is considerable public demand for 3D content, but special glasses are needed, which limits their appeal. 3D4YOU will address the consumer market that coexists with digital cinema and computer games. The 3D4YOU project aims to pave the way for the introduction of a 3D TV system. The project will build on previous European research on 3D, such as the FP5 project ATTEST that has enabled European organizations to become leaders in this field. 3D4YOU will maintain the momentum built up and will capitalize on the wealth of experience of the participants in this project. Combining strengths from different European countries, the 3D4YOU consortium covers all important aspects of the 3D broadcast chain. Its objective is to bring these experts together to deliver an end-to-end system for 3D high-quality media. The generated knowledge will have sustainable effects on the European community. On one hand, it will strengthen the existing expertise in program making and will make sure that the media industry gains technical advances to compete with the worldwide competition. On the other hand, it will enable the European industry to acquire advanced knowledge to develop new products for the emerging market of 3DTV systems.

3D4YOU endeavors to establish practical 3DTV. The key success factor is 3D content. The project seeks to define a 3D delivery format and a content creation process. Establishing practical 3DTV will then be demonstrated by embedding this content creation process into a 3DTV production and delivery chain, including capture, image processing, delivery, and then display in the home. A key project will adapt and improve on these elements of the chain so that every part integrates into a coherent interoperable delivery system. The project's objective is to provide a 3D content format that is independent of display technology, and backward compatible with 2D broadcasting. 3D images will be commonplace in mass communication in the near future. Also, several major consumer electronics companies have made demonstrations of 3DTV displays that could be on the market within two years. The public's potential interest in 3DTV can be seen for the success of 3D movies in recent years. 3D imaging is already present in many graphics applications (architecture, mechanical design, games, cartoons, and special effects for TV and movie production).

In the recent years, multiview display technologies have appeared that improve the immersive experience of 3D imaging, which leads to the vision that 3DTV or similar services might become a reality in the near future. In the United States, the number of 3D-enabled digital cinemas is rapidly growing. By 2011, about 4300 theaters are expected to be equipped with 3D digital projectors, with the number increasing every month. Also in Europe, the number of 3D theaters is growing. In Germany, for example, two cinemas (Munich and Nuremberg) are using the RealD system for the presentation of current 3D Hollywood productions such as *Chicken Little*, *Nightmare Before Christmas*, and *Meet the Robinsons*. Two additional digital 3D cinemas opened up at the end of 2007 in Weimar and Dresden (Cinemagnum 3D), showing 3D Hollywood productions as well as theme park 3D films. Availability of content drives this process: *Chicken Little* from Disney, with its experimental release of the film in digital 3D, increased the number of projectors using the 3D format. Several digital 3D films will surface in the months and years to come and several prominent filmmakers have committed to making their next productions in stereo 3D. The movie industry creates a platform for 3D movies, but there is no established solution to bring these movies to domestic market. Therefore, the next challenge is to bring these 3D productions to the living room. 2D to 3D conversion and a flexible 3D format are an important strategic area. It has been recognized that multiview video is a key technology that serves a wide variety of applications, including free-viewpoint and 3D video applications for the home entertainment and surveillance business fields. Multiview video coding and transmission systems are most likely to form the basis for the next generation of TV broadcasting applications and facilities. Multiview video will greatly improve the efficiency of current video-coding solutions performing simulcasts of independent views. The 3D4YOU project builds on the wealth of experience of the major players in European 3DTV and intends to bring the date of the start of 3D broadcasting a step closer by combining their expertise to define a 3D delivery format and a content creation process.

The key technical problems that currently hamper the introduction of 3DTV to the mass market are:

It is difficult to capture 3D video directly using the current camera technology. At least two cameras need to operate simultaneously with an adjustable but known geometry. The offset of stereo cameras needs to be adjustable to capture depth both close by and far away.

Stereo video (acquired with two cameras) is currently not sufficient input for glasses-free, multiview autostereoscopic displays. The required processing such as disparity estimation is noise sensitive, resulting in low 3D picture quality.

3D postproduction methods and 3D video standards are largely absent or immature.

The 3D4YOU project will tackle these three problems. For instance, a creative combination of two or three high-resolution video cameras with one or two low-resolution depth-range sensors may make it possible to create 3D video of good quality without the need for an excessive investment in equipment. This is in contrast to installing, say, 100 cameras for acquisition where the expense may hamper the introduction of such a system.

Developing tools for conversion of 3D formats will stimulate content creation companies to produce 3D video content at acceptable cost. The cost at which 3D video should be produced for commercial operation is not yet known. However, currently, 3D video production requires almost per-frame user interaction in the video, which is certainly unacceptable. This immediately points out the issue that needs to be solved: Currently fully automated generation of high-3D video is difficult; in the future it needs to be fully or at least semi-automatic with an acceptable minimum of manual supervision during postproduction in order to fulfill the expected demands for video content. 3D4YOU will research how to convert 3D content into a 3D broadcasting format and prove the viability of the format in production and over broadcast chains.

Once 3D video production becomes commercially attractive because acquisition techniques and standards are mature, it will impact the activities of content producers, broadcasters, and telecom companies. As a result one may see these companies adopt new techniques for video production just because the output needs to be in 3D. Also, new companies could be founded that focus on acquiring 3D video and preparing it for postproduction. Here there is room for differentiation since, for instance, the acquisition of a sport event will require large baselines between cameras and real-time transmission whereas the shooting of narrative stories will require both small and large baselines and allows some manual postproduction for achieving optimal quality. These activities will require new equipment (or a creative combination of existing equipment) and new expertise.

3D4YOU will develop practical multiview and depth capture techniques. Currently, the stereo video format is the *de facto* 3D standard that is used by the cinemas. Stereo acquisition may for this reason become widespread as an acquisition technique. Cinemas operate with glasses-based systems and can therefore use a theater-specific stereo format. This is not the case for the glasses-free autostereoscopic 3DTV that 3D4YOU foresees for the home. To allow glasses-free viewing with multiple people at home, a wide baseline is needed to cover the total range of viewing angles. The current stereo video that is intended for the cinema will need considerable postproduction to be suitable for viewing on a multiview autostereoscopic display. Producing visual content will therefore become more complex and may provide new opportunities for companies currently active in 3D movie postproduction. According to the NEM Strategic Research Agenda, multiview coding will form the basis for next-generation TV broadcast applications. Multiview video has the advantage that it can serve different purposes. On the one hand the multiview input can be used for 3DTV; on the other hand it can be shown on a normal TV where the viewer can select his or her preferred viewpoint of the action. Of course, a combination is possible where the viewer selects his or her preferred viewpoint on a 3DTV. However, multiview acquisition with 30 views, for example, will require 30 cameras to operate simultaneously. This requires initially a large investment. 3D4YOU therefore sees a gradual transition from stereo capture to systems with many views. 3D4YOU will investigate a mixture of 3D video-acquisition techniques to produce an extended center view plus depth format (possibly with one or two extra views) that is, in principle, easier to produce, edit, and distribute. The success of such a simpler format relies on the ease (read, cost) at which it can be produced. One can conclude that the introduction of 3DTV to the mass market is hampered by (1) the lack of high quality 3D video content, (2) the lack of suitable 3D formats, and (3) the lack of appropriate format conversion techniques. The variety of new distribution media further complicates this.

Hence one can identify the following major challenges that are expected to be overcome by the project:

- **Video acquisition for 3D content.** Here the practicalities of multiview and depth capture techniques are of primary importance; the challenge is to find the tradeoff, such as number of views to be recorded and how to optimally integrate depth capture with multiview. A further challenge is to define which shooting styles are most appropriate.
- **Conversion of captured multiview video to a 3D broadcasting format.** The captured format needs new postproduction tools (such as enhancement and regularization of depth maps or editing, mixing, fading, and compositing of video-plus-depth representations from different sources) and a conversion step generating a suitable transmission format that is compatible with used postproduction formats before the 3D content can be broadcast and displayed.

- **Coding schemes for compression and transmission.** A last challenge is to provide suitable coding schemes for compression and transmission that are based on the 3D broadcasting format under study and to demonstrate their feasibility in field trials under real distribution conditions.

By addressing these three challenges from an end-to-end system point of view, the 3D4YOU project aims to pave the way to the definition of a 3DTV system suitable for a series of applications. Different requirements could be set depending on the application, but the basic underlying technologies (capture, format, and encoding) should maintain as much commonality as possible to favor the emergence of an industry based on those technologies.

6.5.8 3DPresence

The 3DPresence project plans to implement a multiparty, high-end 3D videoconferencing concept that addresses the problem of transmitting the feeling of physical presence in real time to multiple remote locations.* We mentioned 3D teleconferencing in passing in Chapter 1.

At its core, collaboration in most business endeavors is ultimately about presence: the face-to-face contact that is needed to establish a human atmosphere that increases the team's performance, mutual understanding, and trust. Traditional set-top camera videoconferencing systems have failed to meet the "telepresence challenge" of providing a viable alternative for physical business travel that is nowadays characterized by unacceptable delays, costs, inconvenience, and an increasingly large ecological footprint. In order to provide an experience of presence in an audio/visual conference, it is important to identify first the key physical conditions that give rise to the sensation of "being there." 3DPresence is committed to identifying those requirements and to design the technologies that precisely simulate the spatiotemporal conditions of the real world. In a real presence communication system, the parties should not be able to perceive any middle technology assisting the communication. That is, a conferee should perceive the real geometrical properties of the rest of conferees in a transparent

---

* This section is based on reference [3DR200901]

way, without needing to do any interpretation of the displayed views. For this reason, a true presence communication system must utilize displays that are able to reproduce the size, color, and depth of the transmitted three-dimensional scene, and should also include tracking and rendering technologies capable of mapping the real world from one end point to the specific perspective that the rest of the conferees expect. In short, 3D display technologies and nonverbal communication delivery techniques will have to be combined in the video conferencing arena. The development from 2D toward 3D audio/visual communication is one of the key components for the envisioned system because a true telepresence system needs 3D displays for realistic rendering of the remote participants in addition to a mechanism to keep eye contact and reproduce the correct gesture perspectives, both heavily used subconsciously as part of nonverbal communication.

Traditional videoconferencing systems still fail to meet the telepresence challenge of providing a viable alternative for physical business travel that is currently characterized by unacceptable delays, costs, inconvenience, and an increasingly large ecological footprint. Even recent high-end commercial solutions such as Cisco's TelePresence, Polycom's TPX, and HP's Halo, while partially removing some of these traditional shortcomings, still present the problems of not scaling easily, are relatively expensive to implement, and do not utilize 3D life-sized representations of the remote participants. As a result, none of them is able to convey a natural impression to the remote conferees. One of the fundamental problems with many of the current systems is that eye contact is unnatural and directional gaze awareness is missing. Keeping eye contact is indeed one of the most relevant and challenging requirements in a telepresence system from a nonverbal communication point of view, and while many attempts have been made, it has not yet been satisfactorily solved. Current state-of-the-art systems address it by mounting the camera behind a semitransparent viewing display, but this common approach is often limited to the special case of having one single conferee at each side of the conference. Further, this approach requires a bulky optical and mechanical mounting that is only acceptable for niche market applications.

The 3DPresence project proposes a research and development agenda that is both timely and necessary. It is born from the realization that effective communication and collaboration with geographically dispersed co-workers, partners, and customers requires a natural, comfortable, and easy-to-use experience that utilizes the full bandwidth of nonverbal communication. With this goal in mind, the 3DPresence project intends to implement a multiparty, high-end 3D videoconferencing concept that will tackle the problem of transmitting the feeling of physical presence in real time to multiple remote locations in a transparent and natural way. In order to realize this objective, 3DPresence will go beyond the current state of the art by emphasizing the transmission, efficient coding, and accurate representation of physical presence cues, such as multiple user autostereopsis, multiparty eye contact, and multiparty gesture-based interaction.

The major challenge of the 3DPresence project is to maintain eye contact, gesture awareness, 3D life-sized representations of the remote participants, and the feeling of physical presence in a multiparty, multiuser terminal conference system. In order to achieve these objectives, the concept of a shared virtual table is applied. All remote conferees will be rendered based on a predefined shared virtual environment. Eye contact and gesture awareness can be created by adapting virtually the 3D perspective and 3D position of all remote conferees on each of the terminal displays. Furthermore, in order to maximize the feeling of physical presence, sophisticated multiuser 3D display technologies will be developed and applied within the 3DPresence project (for example, see Figure 6.1). The concept will be proved by

Figure 6.1 (Please see color insert following page 160) 3DPresence multiparty video-conferencing concept. (From 3DPresence Project materials.)

Figure 6.2 (Please see color insert) Drawing of the mock-up system. (From 3DPresence Project materials.)

developing a real-time demonstrator prototype system consisting of three 3D videoconferencing stations in different European countries.

Current 3D displays are capable of providing multiple stereoscopic views in order to support head motion parallax. Hence the user is able to perceive a scene in 3D and recognize different perspectives according to the user's head position. In contrast, the challenge in 3DPresence is to provide stereoscopic viewing for two users and head motion parallax with significantly different perspectives onto the scene. The envisaged approach is to develop a novel multiview 3D display that provides two viewing cones providing significantly different perspectives with each of them supporting multiple views. Due to this novel display design, the viewing cones of four displays, related to the four remote conferees, must meet at the correct position of the two local conferees. This has significant impact on the overall design of the telepresence system as the overlapping area of all viewing cones related to one local conferee must be as large as possible. Based on an initial study in the 3DPresence project, a mock-up system has been installed in order to test different camera and display configurations as well as to capture test sequences for algorithm development. Figure 6.2 depicts a mock-up of a system being studied by the 3DPresence (this is a CAD-based simulation of the proposed system under study).

6.5.9 Audio-Visual Content Search and Retrieval in a Distributed P2P Repository (Victory)

Victory is a European Commission DG Information Society and Media FP6–funded research project that will develop an innovative, distributed search engine, introducing MultiPedia search and retrieval capabilities to a standard (PC-based) and a mobile peer-to-peer (P2P) network.* A MultiPedia object is defined as a 3D object along with its accompanied information, that is, 2D views, text, audio, video. Goals of the project include:

- Development of the first distributed 3D search engine that will enable searching, accessing, and retrieving 3D and other types of digital content in a distributed MultiPedia object repository, through peer-to-peer networks
- Development of innovative 3D search and retrieval techniques supporting mixed-media searching based on the extraction of low-level geometric characteristics (content) and intuitive semantics (context) from the audiovisual content
- Development of an open-source, self-contained, P2P-based middleware framework that will integrate quality of experience coordination services, identity management services, and knowledge infrastructure services, along with enhancements and extensions to existing distributed-computing technologies (JXTA, OMG DDS, etc.)
- Development of the knowledge infrastructure serving the organization and exploitation of data pertaining to business rules and security policies with regard to 3D content access and manipulation
- Integration of novel mobile interfaces with knowledge extraction algorithms in order to provide next-generation information query and access mechanisms
- Personalization of information access and knowledge-based content manipulation and visualization
- Development of a specialized server network or remote visualization of heavyweight 3D content by low-power mobile devices
- Integration of novel 3D watermarking techniques with state-of-the-art copyright protection techniques

Project objectives are as follows:

- To develop an innovative, distributed search engine that will introduce mixed-media (MultiPedia) search and retrieval

---

* This section is based on reference [VIC200901].

capabilities to a standard (PC-based) and a mobile P2P network. The 3D search engine will be based on:

- Content, which will be extracted from low-level geometric characteristics
- Context, which will be high-level features (semantic concepts) mapped to low-level features.

- To provide a solution so as to bridge the gap between low- and high-level information through automated knowledge discovery and extraction mechanisms. High-level features will be based on:
  - Appropriate annotation mechanisms provided by the system or generated by the user dynamically
  - Relevance feedback, where the user marks which retrieved objects were relevant to the query (user subjectivity)

- To deal with the problems raised from the visualization and rendering of "heavyweight" 3D data on mobile devices. Graphics workstations of specialized cluster peers are envisioned to undertake this load, taking advantage of the shared computational power of each peer.

6.5.9.1 Victory in Automotive Industry   In the automotive industry, the use of existing 3D computer-aided design (CAD) data outside the core development process is problematic, especially in the case of mobile platforms. This problem is directly related to the complexity and diversity of the data models, as well as the lack of 3D simplification, sharing, and multimodal delivery tools. Centralized product lifecycle management (PLM) systems are totally dependent on specific vendors. Moreover they do not provide any tools or support for mobile collaborative working. Within the Victory concept, facile and expedited decision making within knowledge-based automotive engineering departments will result from P2P collaborative tools. P2P design and design review networks empower teams and individuals to create working/decision-making groups easily and on the fly, with the focused intent of sharing information across platforms, subsidiaries, subcontractors, and related organizations. Victory technologies and tools will be tested as next-generation product design data management tools in the automotive industry. Automotive designs have been chosen as the best "proof of concept" for the Victory technology due to their high degree of complexity and the diversity of available formats and tools. The Victory tools are expected to enable the reuse of complex engineering designs within the postproduction phase of a product for reengineering, maintenance, education, and marketing purposes.

6.5.9.2 Victory in Game Industry The popularity of massively multiplayer online games is increasing at a steady pace. These games allow the user to customize a multitude of objects of the game world. There are online repositories on the Internet that offer relevant 3D content, either for free or for a small fee. In such a repository, the user can search for content using text-based search and also browse through categories of 3D objects. These online repositories add value to a game, whether this game is commercial or open source. What Victory could add to those repositories is the ability to search, based on 3D similarity and context, and also share 3D content among users of those communicating using the P2P interface proposed. The Victory framework could also contribute to the customization and added value of 3D video games, both on home entertainment systems (such as personal computers and video game consoles) and on emerging mobile entertainment platforms (like mobile phones, handheld videogames, and PDAs). The use of Victory would facilitate the retrieval and the exchange of game avatars, objects, levels, and in general any 3D object that could be added in a game.

6.5.10 2020 3D Media

2020 3D Media has the goal to research, develop, and demonstrate novel forms of compelling entertainment experiences based on new technologies for the capture, production, networked distribution, and display of three-dimensional sound and images.*

Specifically, the goal is to explore and develop new technologies to support the acquisition, coding, editing, networked distribution, and display of stereoscopic and immersive audio/visual content providing novel forms of compelling entertainment at home or in public spaces. The users of the resulting technologies will be both media industry

---

* This section is based on reference [3D2200901].

professionals across the current film, TV, and new media sectors producing program material and the general public.

The media industry knows that astonishing the public is still a route to large audiences and financial success. It is believed that high-quality presentation of stereoscopic or immersive images in the home and in public entertainment spaces (such as cinemas) can offer previously unimagined levels of experience. The potential advantages of spatialized stereoscopic or immersive entertainment systems include:

- Heightened reality and a renewed sense of presence, putting the spectator at the heart of a more exciting experience
- The ability for the spectator to navigate a virtualized world that has a complete sense of reality
- The ability to change things in this world once it has been created
- The ability to repurpose and deploy multidimensional content in different contexts

The market for entertainment products continually evolves in the search for ways to improve the audience experience and sense of involvement. 2D representations no longer give the sensation of presence (even with widescreen and Dolby sound). There are different approaches to increasing the sense of presence by using the third dimension: one involves stereoscopic display, giving the illusion of three-dimensionality to the scene; another is to immerse the viewer in a three-dimensional environment. Both these approaches are difficult and neither has yet been widely adopted by the media industry. Most "3D media" currently exhibited is artificially processed 2D material and the dimensionalization process gives varying quality and depth representation: It is hard to get beyond the sensation of rather flat objects with depth separation. The special spectacles needed to view most current 3D products are uncomfortable and cause headaches (which are not experienced by people viewing real 3D scenes). They also only give a true perspective vision for one viewer location. Despite 3D computer-generated imagery (CGI) being widely used for films shown in 2D, virtual reality and immersive methods have so far only found a limited application in simulation-based entertainment.

The project will result in new compelling and thrilling entertainment experiences. 2020 3D Media will demonstrate and evaluate experimental productions along with the enabling hardware and software technology. Standardization initiatives will help to establish an open (nonproprietary) data format and process framework for industry-wide development of spatial media and media technologies, with application potential further than purely leisure functions.

The innovation from the 2020 3D Media project consists of new and engaging 3D media forms, creating the technologies for producing and presenting these 3D-surround audio/visual media as a viable product. Through the project, an end-to-end system will be developed and demonstrated, consisting of capture, postproduction, secure network transmission, play-out, and end-user customization blocks. Both media and associated technologies are intended to possess a compelling nature, with high degrees of realism and thrill. One of the fundamental aims of the project is to recapture the excitement once associated with cinema in its pioneering days, and to extend it over diverse forms of entertainment, including home cinema and immersive interactive media products.

Key factors required to achieve this degree of innovation are:

- Technologies and formats for 3D-sound and 3D-image capture and coding, including novel high-resolution cameras
- Technologies and methods for 3D sound and image postproduction
- Technologies for the secure distribution and display of the new media
- Tools for the new spatial media technologies and their creative application

2020 3D Media expects to produce a significant scientific, technical, and socioeconomic impact that can be summarized through the following achievements:

- World leadership in a new generation of media technologies, providing significantly higher performances in terms of built-in intelligence, scalability, flexibility, speed, capacity, ease of use, and cost.
- Quick response to new and sustainable market opportunities based on converged business models between content, telecom, broadcast, and consumer electronics industries

- Widespread adoption of new digital media consumption and production patterns
- Enhanced quality of life through new usage forms contributing to social, intellectual, and leisure well-being
- New opportunities for content production and exploitation

6.5.11 i3DPost

i3DPost is a European Union Project under Framework 7 ICT Programme. This project is formed of six companies and educational institutions that are Europe's leading specialists in their fields. The project was expected to run for three years. i3DPost will develop new methods and intelligent technologies for the extraction of structured 3D content models from video at a level of quality suitable for use in digital cinema and interactive games. The research will enable the increasingly automatic manipulation and reuse of characters, with changes of viewpoint and lighting. i3DPost will combine advances in 3D data capture, 3D motion estimation, postproduction tools, and media semantics. The result will be film-quality 3D content in a structured form, with semantic tagging, that can be manipulated in a graphic production pipeline and reused across different media platforms [I3D200901].

References

[3D2200901] 2020 3D Media Project, 3D Media Cluster, Umbrella structure embracing related EC 3DTV funded projects, http://www.20203dmedia.eu/

[3D4200901] 3D4YOU Project, 3D Media Cluster, Umbrella structure embracing related EC 3DTV funded projects, http://www.3d4you.eu/index.php

[3DA201001] The 3D@Home Consortium. http://www.3dathome.org/

[3DM201001] The 3D Media Cluster, http://www.3dmedia-cluster.eu/

[3MO200901] Mobile3DTV Project, 3D Media Cluster, Umbrella structure embracing related EC 3DTV funded projects, http://sp.cs.tut.fi/mobile3dtv

[3DP200901] 3DPhone Project, 3D Media Cluster, Umbrella structure embracing related EC 3DTV funded projects, http://the3dphone.eu/

[3DP200801] 3DPhone, Project no. FP7–213349, Project title All 3D Imaging Phone, 7th Framework Programme, Specific Programme "Cooperation," FP7-ICT-2007.1.5—Networked Media, D5.2—Report on first study results for 3D video solutions, 31 December 2008.

[3DR200901] 3DPresence Project, 3D Media Cluster, Umbrella structure embracing related EC 3DTV funded projects, http://www.3dpresence. eu/

[3DT200401] IST–6th Framework Programme, 3DTV NoE, L. Onural, EEE Department, Bilkent University, TR-06800 Ankara, Turkey.

[3DT200601] IST–6th Framework Programme, 3DTV NoE, Mid Course report, 2006, EEE Department, Bilkent University, TR-06800 Ankara, Turkey.

[3DT200801] 3DTV, 3DTV Project, 3D Media Cluster, Umbrella structure embracing related EC 3DTV funded projects, http://www.3dtv-research. org/

[FEH200401] C. Fehn, Depth-image-based rendering (DIBR), compression, and transmission for a new approach on 3DTV, in *Stereoscopic Displays and Virtual Reality Systems XI.* A.J. Woods, J.O. Merritt, S.A. Benton, M.T. Bolas, Eds., Proceedings of the SPIE, 5291, 93–104, 2004.

[HEL200901] HELIUM3D Project, 3D Media Cluster, Umbrella structure embracing related EC 3DTV funded projects, http://www.helium3d.eu/

[I3D200901] i3DPost Project, 3D Media Cluster, Umbrella structure embracing related EC 3DTV funded projects, http://www.i3dpost.eu/

[MUT200901] MUTED Project, 3D Media Cluster, Umbrella structure embracing related EC 3DTV funded projects, http://www.muted3d.eu/

[REA200901] Real3D Project, 3D Media Cluster, Umbrella structure embracing related EC 3DTV funded projects, http://www.digitalholography. eu/

[VIC200901] Victory Project, 3D Media Cluster, Umbrella structure embracing related EC 3DTV funded projects, http://www.victory-eu.org:8080/ victory.

Glossary[*]

1080p: 1080p is a high-definition video format with resolution of 1920 × 1080 pixels. The "p" stands for progressive scan, which means that each video frame is transmitted as a whole in a single sweep. The main advantage of 1080p TVs is that they can display all high-definition video formats without down-converting, which sacrifices some picture detail. 1080p TVs display video at 60 frames per second (fps), so this format is often referred to as 1080p60. The video on most high-definition discs is encoded at film's native rate of 24 fps, or 1080p24. For compatibility with most current 1080p TVs, high-definition players internally convert the 1080p24 video to 1080p60. Newer TVs have the ability to accept a 1080p24 signal directly. These TVs do not actually display video at 24 fps because that would cause visible flicker and motion stutter. The TV converts the video to 60 fps or whatever its native display rate is. The ideal situation would be to display 1080p24 at a multiple of 24 fps, such as 72, 96, or 120 fps, to avoid the motion shudder caused by 3-2 pull-down that is required when converting 24 fps material to 60 fps [KIN200901].

---

[*] Portions of this Glossary are provided by the 3D@Home Consortium and used with permission [3DA201001]. Other material added by author.

120 Hz refresh rate: The digital display technologies (LCD, plasma, DLP, LCoS, etc.) that have replaced picture tubes are progressive scan by nature, displaying 60 video fps—often referred to as "60 Hz." HDTVs with 120 Hz refresh rate employ sophisticated video processing to double the standard rate to 120 fps by inserting either additional video frames or black frames. Because each video frame appears for only half the normal amount of time, on-screen motion looks smoother and more fluid, with less smearing. It is especially noticeable viewing fast-action sports and video games. This feature is available on an increasing number of flat-panel LCD TVs [KIN200901].

240 Hz refresh rate: 240 Hz refresh rate reduces LCD motion blur even more than 120 Hz refresh rate. 240 Hz processing creates and inserts three new video frames for every original frame. Most "240 Hz" TVs operate this way, but some models use "pseudo-240 Hz" technology that combines 120 Hz refresh rate with high-speed backlight scanning. An example of the pseudo-240 Hz approach that is very effective is Toshiba's ClearScan 240 technology [KIN200901].

2D: Two dimensional. An image or object with only two dimensions, such as width and height, but no depth.

2D+delta: A single image along with data that represents the difference between that image view and a second eye image view along with other additional metadata. The delta data could be spatiotemporal stereo disparity, temporal predictive, or bidirectional motion compensation.

2D signal processing: A signal-processing chain where 2D and 3D signals receive the same processing steps and the processor does not need to know what type of signal is being processed.

3D: Having or appearing to have width, height, and depth (three dimensional). Accepts and/or produces uncompressed video signals that convey 3D.

3D adjustment setting: Changes the apparent depth of objects on a 3D view screen.

3D distribution (or transport) formats: Formats for 3D content transmitted to the end user over the air, over cable, over satellite, over the Internet, or on packaged media. These formats typically need to be compressed on the service provider side and decompressed on the network termination at home.

3D DVD: A DVD movie recorded in 3D

3D format: An uncompressed video signal type used to convey 3D over an interface.

3D in-home formats: Formats used when connecting in-home devices to the 3D display system. In-home formats may be compressed or uncompressed. The decompression and decoding/transcoding can be done in several places in the home and can include additional demodulation of radio frequency (RF) modulated signals as well. Video decoding and 3D decoding may be done at different locations in the signal chain, which could require two different in-home formats.

3D native display formats: Formats that are required to create the 3D image on the particular TV. These formats may reside only in the TV or can be decoded/transcoded outside of the TV. Normally, once a signal is decoded into the 3D native display format, no additional 3D signal processing is required to display the signal, although there is likely to be additional 2D processing. The 3D native display format is different from the native 3D display format or resolution, which refers to the 3D pixel arrangement.

3D ready: Contains 3D decoder/transcoder and may accept and produce uncompressed video signals that convey 3D.

3D reconstruction: The generation of a portion of a scene in the presence of occlusion; plays a role in multiview TV. Methods include a continuous geometric formulation solved by local optimization methods such as *space carving* or *level sets* and a discrete labeling formulation for computing stereo disparities solved by global *graph-cut* methods. Direct volumetric methods: The method of *space carving* or *voxel coloring* directly works on discretized 3D space (voxels) based on their image consistency and visibility. These techniques are characterized by their local treatment; voxels are considered separately. Volumetric method based on level sets: A variational approach implemented by level sets. This approach naturally handles the changes in topology and occlusion problems, but it is not clear under what conditions this method converges as the actual proposed functional seems highly nonconvex. Disparity methods based on graph cuts: Graph-cut methods are an extension of dynamic programming. This can be interpreted as a labeling problem in the Markov random field framework. Functionals with various concave smoothing terms have been introduced but the resulting problem is unfortunately nondeterministic polynomial-time (NP-hard). Geometric methods based on graph cuts: A geometric formulation where one computes geodesic surfaces for data segmentation [COS200701].

3D rendering: The process of producing an image based on three-dimensional data stored within a computer.

3D signal processing: A video signal-processing chain where the processing of the signal is different for 3D video than it is for 2D video and the processor must be aware of the type of signal it is processing.

3D video (3DV): Any capture, transmission, or display of three-dimensional video content in any venue, including playback of downloaded and stored content, display of broadcasted content (i.e., 3DTV, via DVB, DVB-H), display of streamed content (i.e., via mobile phone line, WLAN), or recording.

3D viewing: The act of viewing a 3D image with both eyes in order to experience stereoscopic vision and binocular depth perception.

Accommodation: The refocusing of the eyes as their vision shifts from one distance plane to another.

Accommodation-convergence link: The physiological link that causes the eyes to change focus as they change convergence, a link that has to be overcome in stereo viewing since the focus remains unchanged on the plane of the constituent flat images.

Accommodation-convergence relationship: The learned relationship established through early experience between the focusing of the eyes and verging of the eyes when looking at a particular object point in the visual world; usually called the accommodation-convergence relationship (or the convergence-accommodation relationship).

Accommodative facility: The eyes' ability to repeatedly change focus from one distance to another, often measured by use of special flipper lenses. Measurement of each eye in turn is usually made followed by comparing the performance to that of both eyes working together.

Active glasses: Powered shutter glasses that function by alternately allowing each eye to see the left-eye/right-eye images in an eye-sequential 3D system. Most commonly based on liquid crystal devices. *See* Passive glasses.

Active stereo: *See* Eye-sequential 3D.

Addressable hologram: A hologram that can be changed in real time or near real time.

AIP (anterior intraparietal cortex): An area of the human brain that is uniquely sensitive to visual cues.

ALiS (alternate lighting of surfaces): A type of high-definition plasma TV panel designed for optimum performance when displaying 1080i material. On a typical progressive-scan plasma TV, all pixels can be illuminated at any given instant. With an ALiS plasma panel, alternate rows of pixels are illuminated so that only half the panel's pixels can be illuminated at any moment (somewhat similar to interlaced-scanning on a CRT-type TV). ALiS-based plasmas make up a small part of the overall market; TV makers that use ALiS panels include Hitachi and Fujitsu [KIN200901].

Amblyopia: "Lazy eye"; a visual defect that affects approximately 2 or 3 out of every 100 children in the United States. Amblyopia involves lowered visual acuity (clarity) and/or poor muscle control in one eye. The result is often a loss of stereoscopic vision and binocular depth perception.

Anaglyph: A type of stereogram (either printed, projected, or viewed on a TV or computer screen) in which the two images are superimposed but are separated, so each eye sees only the desired image by the use of colored filters and viewing spectacles (commonly red and cyan, or red and green). To the naked eye, the image looks overlapping, doubled, and blurry. Traditionally, the image for the left eye is printed in red ink and the right-eye image is printed in green ink.

Analyzers: Devices placed in front of the eyes to separate the left- and right-eye images, mainly when projected. Typically, these are polarizing spectacles, anaglyph spectacles, or liquid-crystal shutters.

Anamorphic video: Refers to widescreen video images that have been "squeezed" to fit a narrower video frame when stored on DVD. These images must be expanded (unsqueezed) by the display device. Most of today's TVs employ a screen with 16:9 aspect ratio, so that anamorphic and other widescreen material can be viewed in its proper proportions. When anamorphic video is displayed on an old-fashioned TV with a 4:3 screen, images appear unnaturally tall and narrow [KIN200901].

Angular resolution: The angular resolution determines the smallest angle between independently emitted light beams from a single screen point. It can be calculated by dividing the emission range with the number of independently addressable light beams emitted from a screen point. The angular resolution determines the smallest feature (voxel) the display can reconstruct in a given distance from the screen.

Aspect ratio: The ratio of width to height for an image or screen. The North American NTSC television standard uses the squarish 4:3 (1.33:1) ratio. HDTVs use the wider 16:9 ratio (1.78:1) to better display widescreen material such as high-definition broadcasts and DVDs [KIN200901].

Autostereoscopic: 3D displays that do not require glasses to see the stereoscopic image. Multiview autostereoscopic displays based on parallax barriers or lenticules are sometimes called parallax panoramagram displays.

Backlight scanning: An anti-blur technology used in some LCD TVs. Typical LCDs use a fluorescent backlight that shines constantly, which can contribute to motion blur. LCD models with backlight scanning have a special type of fluorescent backlight that pulses at very high speed, which has the effect of reducing motion blur. Some recent TVs use backlight scanning along with 120 Hz refresh rate for even greater blur reduction [KIN200901].

Beam splitter: A device consisting of prisms and/or mirrors that can be attached to a mono camera to produce two side-by-side images (usually within a single frame). (More accurately described as an image splitter, as it does not split an individual beam into components.) Because the groups of light rays forming the left and right images cross over as they pass through the camera lens, the recorded images end up in the correct configuration for stereo viewing without the need for the usual transposition.

Binocular: Of or involving both eyes at once. The term *binocular stereopsis* (two-eyed solid seeing) is used in some psychology books for the depth sense more simply described as stereopsis.

Binocular depth perception: A result of successful stereo vision; the ability to visually perceive three-dimensional space; the ability to visually judge relative distances between objects; a visual skill that aids accurate movement in three-dimensional space.

Binocular disparity: The difference between the view from the left and right eyes.

Binocular vision: Vision as a result of both eyes working as a team; when both eyes work together smoothly, accurately, equally, and simultaneously.

Binocular vision disability: A visual defect in which the two eyes fail to work together as a coordinated team resulting in a partial or total loss of binocular depth perception and stereoscopic vision. At least 12 percent of the population has some type of binocular vision disability. Amblyopia and strabismus are the most commonly known types of binocular vision disabilities.

Breaking the frame: If an object has negative parallax and is bisected by the edge of the frame, then that object is "breaking the frame." This gives rise to a visual/brain conflict.

Broadband light: Light with a range of optical wavelengths that is comparable to the bandwidths associated with the red, green, and blue lights of a display.

Cardboarding: A condition where objects appear as if cut out of cardboard and lack individual solidity; usually the result of inadequate depth resolution arising from, for example, a mismatch between the focal length of the taking lens, the stereo base, and/or the focal length of the viewing system.

Chromatic stereoscopy: An impression of depth that results from viewing a spectrum of colored images through a light-bending device such as a prism, a pinhole, or an embossed "holographic" filter, caused by variations in the amount of bending according to the wavelength of the light from differing colors (chromatic dispersion). If such a device is placed in front of each eye but arranged to shift planar images or displays of differing colors laterally in opposite directions, a 3D effect will be seen. The effect may also be achieved by the lenses of the viewer's eyes themselves when viewing a planar image with strong and differing colors. Typically, with unaided vision, red portions of the image appear closer to the viewer than the blue portions of the image. Sometimes called chromostereopsis.

Circular polarization: A form of polarized light in which the tip of the electric vector of the light ray moves through a corkscrew in space.

Column interleaved format: A 3D image format where left and right view image data are encoded on alternate columns of the display.

Compressed video signal: A stream of compacted data representing an uncompressed video signal. A compressed video signal is an encoded version of an uncompressed video signal. A compressed video signal must be decoded to an uncompressed video signal in order to be edited or displayed. Compressed video formats vary according to the encoding methods used. A compressed video signal format may be converted to another using a "transcoder."

Computer-generated holograms (CGHs): The counterpart of computer graphics imagery (CGI) in holography. The technology has a long history and is sometimes referred to as the final 3D technology, because CGHs not only produce a sensation of depth but also generate light from the objects themselves. However, currently available CGHs cannot yet produce fine, true 3D images accompanied by a strong sensation of depth. Such fine CGHs commonly require the following two conditions: the CGHs must have a large viewing zone to acquire the autostereoscopic property (i.e., motion parallax) and the dimensions of the CGHs must be large enough to reconstruct a 3D object that can be observed by two naked eyes. Both of these conditions lead to an extremely large number of pixels for a CGH because the large viewing-zone requires high spatial resolution and the large dimensions require a large number of pixels for high resolution. In addition, scenes with occlusions should be reconstructed to give a CGH a strong sensation of depth, because the ability to handle occlusions is one of the most important mechanisms in the perception of 3D scenes. The reason why fine 3D images are difficult to be produced by CGH technology is that there is no practical technique to compute the object fields for such high-definition CGHs that reconstruct 3D scenes with occlusions [MAT200901].

Contrast ratio: Measures the difference between the brightest whites and the darkest blacks that a TV can display. The higher the contrast ratio, the better a TV will be at showing subtle color details, and the better it will look in rooms with more ambient room light. Contrast ratio is one of the most important specs for all TV types. There are actually two different ways of measuring a TV's contrast ratio. Static contrast ratio measures the difference between the brightest and darkest images a TV can produce simultaneously (sometimes called on-screen contrast ratio). The ratio of the brightest and darkest images a TV can produce over time is called dynamic contrast ratio. Both specs are meaningful, but the dynamic spec is often four or five times higher than the static spec [KIN200901].

Conventional stereo video (CSV): Conventional stereo video is the most well-known and in a way most simple type of 3D video representation. Only color pixel video data are involved, which are captured by at least two cameras. The resulting video signals may undergo some processing steps like normalization, color correction, rectification, etc., but in contrast to other 3D video formats, no scene geometry information is involved. The video signals are meant in principle to be directly displayed using a 3D display system, though some video processing might also be involved before display.

Convergence: The ability of both eyes to turn inward together. This enables both eyes to be looking at the exact same point in space. This skill is essential to being able to pay adequate attention at near points to be able to read. Not only is convergence essential to maintaining attention and single vision, it is vital to be able to maintain convergence comfortably for long periods of time. For good binocular skills, it is also necessary to be able to look further away. This is called divergence. Sustained ability to make rapid convergence and divergence movements are vital skills for learning. [The term has also been used, confusingly, to describe the movement of left and right image fields or the rotation (toe-in) of camera heads.]

Corresponding points: The image points of the left and right fields referring to the same point on the object. The distance between the corresponding points on the projection screen is defined as parallax. Also known as conjugate or homologous points.

Cross-talk: Incomplete isolation of the left and right image channels so that one leaks or bleeds into the other (leakage); looks like a double exposure. Cross-talk is a physical entity and can be objectively measured, whereas ghosting is a subjective term. *See also* Ghosting.

CRT: Cathode ray tube. Legacy display technology direct view CRTs have often been used in eye-sequential 3D systems but alternative 3D display systems are now common.

Depth budget: The combined values of positive and negative parallax; often given as a percentage of screen width.

Depth-image-based rendering (DIBR): An approach to 3DTV where the input is comprised of a monoscopic stream plus a depth stream, and the output to the display device is a stereoscopic stream. Depth information indicates the distance of an object in the three-dimensional (3D) scene from the camera viewpoint, typically represented by 8 bits.

Depth information: Depth information indicates the distance of an object in the three-dimensional (3D) scene from the camera viewpoint, typically represented with an 8-bit codeword. Depth maps are applicable to a number of multimedia applications, including 3DTV and free-viewpoint television (FTV).

Depth perception: The ability to see in 3D or depth to allow us to judge the relative distances of objects; often referred to as stereo vision or stereopsis.

Depth range: A term that applies to stereoscopic images created with cameras. The limits are defined as the range of distances in camera space from the background point producing maximum acceptable positive parallax to the foreground point producing maximum acceptable negative parallax.

Diplopia: "Double vision." In stereo viewing, a condition where the left and right homologues in a stereogram remain separate instead of being fused into a single image.

Direct view: A display where the viewer looks directly at the display, not at a projected or virtual image produced by the display. CRTs, LCDs, plasma panels, and OLEDs can all be used in direct-view 3D displays.

Discrete views: The 3D view from any position is provided by a single image source. *See also* Distributed views.

Disparate images: A pair of images that fail as a stereogram (e.g., due to distortion, poor trimming, masking, mismatched camera lenses).

Disparity: The distance between conjugate points on overlaid retinas, sometimes called retinal disparity. The corresponding term for the display screen is parallax.

Disparity difference: The parallax between two images representing the same scene but acquired from two different viewing angles. The disparity between homologous points is used to compute the elevation.

Display: An electronic device that presents information in visual form, that is, produces an electronic image—such as CRTs, LCDs, plasma displays, electroluminescent displays, field emission displays, etc. Also known as a "sink" that renders an image.

Display surface: The physical surface of the display that exhibits information (synonym: screen).

Distortion: In general usage, any change in the shape of an image that causes it to differ in appearance from the ideal or perfect form. In stereo, usually applied to an exaggeration or reduction of the front-to-back dimension.

Distributed views: The 3D view at any one time and position from multiple image sources. Also see Discrete views.

Divergence: The ability for the eyes to turn outward together to enable them to both look further away; the opposite of convergence. It is essential for efficient learning and general visual performance to have good divergence and convergence skills.

DLNA (Digital Living Network Alliance): A collaboration among more than 200 companies, including Sony, Panasonic, Samsung, Microsoft, Cisco, Denon, and Yamaha. Their goal is to create products that connect to each other across a home network, regardless of manufacturer, so one can easily enjoy the digital and online content in any room. While all DLNA-compliant devices are essentially guaranteed to work together, they may not be able to share all types of media. For example, a DLNA-certified TV might be able to display digital photos from a DLNA-certified media server but not videos [KIN200901].

DLP (digital light processing): A projection TV technology developed by Texas Instruments, based on their digital micromirror device (DMD) microchip. Each DMD chip has an array of tiny swiveling mirrors that create the image. Depending on the TV's resolution, the number of mirrors can range from several hundred thousand to over two million. DLP technology is used in both front- and rear-projection displays. There are two basic types of DLP projectors. "Single-chip" models, which include virtually all rear-projection DLP TVs, use a single DMD chip, with color provided by a spinning color wheel or colored LEDs. "Three-chip" projectors dedicate a chip each to red, green, and blue. While three-chip models are considerably more expensive, they completely eliminate the rainbow effect, which is an issue for a small minority of viewers [KIN200901].

DVI (digital visual interface): A multipin, computer-style connection intended to carry high-resolution video signals from video source components (such as older HD-capable satellite and cable boxes, and up-converting DVD players) to HD-capable TVs with a compatible connector. Most (but not all) DVI connections use HDCP (high-bandwidth digital content protection) encryption to prevent piracy. In consumer electronics products, DVI connectors have been almost completely replaced by HDMI connectors that carry both video and audio. One can use an adapter to connect a DVI-equipped component to an HDMI-equipped TV, or vice versa, but a DVI connection can never carry audio [KIN200901].

Dwarfism: *See* Lilliputism.

Emissive: A self-luminous display where there is no separate lamp. CRTs, plasma panels, LEDs, and OLEDs are examples.

Extrastereoscopic cues: Those depth cues appreciated by a person using only one eye; also called monocular cues. They include interposition, geometric perspective, motion parallax, aerial perspective, relative size, shading, and textural gradient.

Eye-dedicated displays: A 3D display system where there are two separate displays to produce the left- and right-eye images, and the geometry of the system is arranged so each eye can only see one display.

Eye-sequential 3D: The images in a stereopair are presented alternately to the left and right eyes fast enough to be merged into a single 3D image. At no instant in time are both images present. The images may be separated at the eyes by active or passive glasses.

Eye tracking: *See* Tracking.

Eyewear: Anything worn on the head and eyes to produce a 3D image. This includes both passive and active glasses or head-mounted displays (HMDs). Consumer-grade 2D and 3D HMDs are often specifically called eyewear. Passive and active glasses are often just called glasses.

Far point: The feature in a stereo image that appears to be farthest from the viewer.

Field of depth: The field of depth determines the largest depth a display can visualize with a defined minimum resolution. For displays with fixed emission range and angular resolution, the size of the smallest displayed feature depends on the distance from the screen. The smallest feature (voxel) the display can reconstruct is the function of the distance from the screen and the angular resolution. If one sets an upper limit on the feature size, the angular resolution determines the distance from the screen, within which the displayed features are smaller than the given limit. This range is the field of depth, which effectively determines the largest displayable depth below which the features are within given limit.

Field of view: Usually measured in degrees, this is the angle at which a lens can accept light. For instance, the human eye's horizontal field of view is about 175 degrees.

Field sequential: The rapid alternation of left and right views in the video format, on the display, or at the eye.

Fields per second: The number of subimages presented each second. The subimage can be defined by the interlace pattern, the color, or the left/right images in a stereopair.

Film: A sheet of material that is thin compared to its lateral dimensions. Films are used to modify the light passing through or reflecting off of them. Films can modify the brightness, color, polarization, or direction of light. Film encoded with images can be used in projection systems as an image source.

Flat: *See* Planar image.

Flat-panel display (FPD): The two most common FPDs used in 3D systems are LCDs and plasma panels. OLED FPDs are also becoming commercially available.

Flat-panel TV: Any ultra-thin, relatively lightweight TV, especially those that can be wall mounted. Current flat-panel TVs use either plasma or LCD screen technology.

Floating image: A display where the image appears to be floating in midair, separated from any physical display screen.

Fore window image: An image that appears in front of the stereo window frame, i.e., "coming through the window." Where an image cuts the edge of the window-frame, the effect is usually referred to as floating edges.

Format: The method used to combine images for printing, storage, or transmission.

Frame: In moving picture media, whether film or video, a frame is a complete, individual picture.

Frame-compatible 3D format: Left/right frames organized to fit in a single legacy frame such as 480 × 720, 720 × 1280, or 1080 × 1920 pixels. The pair of images can be pixel decimated using spatial compression, color encoded like anaglyph, time sequenced like page flipping, etc.

Frame rate: The rate at which frames are displayed. The frame rate for movies on film is 24 frames per second (24 fps). Standard NTSC video has a frame rate of 30 fps (actually 60 fields per second). The frame rate of a progressive-scan video format is twice that of an interlaced-scan format. For example, interlaced formats like 480i and 1080i deliver 30 complete frames per second; progressive formats like 480p, 720p, and 1080p provide 60 [KIN200901].

Frames per second (fps): The number of complete images delivered to the eye each second.

Free-viewpoint television (FTV): Free-viewpoint TV/video is a viewing environment where the user can choose his or her own viewpoint. Video arrangement where the user can choose his or her own viewpoint require a 3D video format that allows rendering a continuum of output views or a very large number of different output views at the decoder.

Free-viewpoint video (FVV): *See* Free-viewpoint television.

Front-projection TV: *See* Projector.

Frustum: The frustum is the rectangular wedge that one gets if a line is drawn from the eye to each corner of the projection plane, for example, the screen.

Frustum effect: Front-to-back keystone distortion in the space-image so that a cube parallel to the lens base is portrayed as the frustum of a regular four-sided, truncated pyramid with the smaller face toward the observer. In reverse frustum distortion, the larger face is forward.

Full-frame stereo format: A stereo format that uses stereopairs of eight perforations (film sprockets) per image width. This would be the same as a conventional camera and is used on twin camera stereo photographs and with certain RBT cameras.

Fusion: The merging (by the action of the brain) of the two separate views of a stereopair into a single three-dimensional (or Cyclopean) image.

Fusion, irregular: Fusion of points that are not homologous, as with accidental and false stereo effects and multiple diplopia.

Fusional reserves: A series of measures to probe how much stress the convergence and divergence mechanisms are able to cope with when placed under stress. This is linked to the ability to maintain good, clear, comfortable single vision while keeping control of the focusing mechanism. Analysis of the results of this test is complicated. If results are low, it can be expected that difficulty in concentrating for long periods will be experienced. Often headaches can result in prolonged periods of close work. Children in particular, but also adults, often show a tendency to avoid prolonged close work when the fusional reserves are low.

Ghosting: The perception of cross-talk is called ghosting, a condition that occurs when the right eye sees a portion of the left image or vice versa, causing a faint double image to appear on the screen.

Giantism: Jargon term for the impression of enlarged size of objects in a stereo image due to the use of a stereo base separation less than normal for the focal length of the taking lens(es). *See also* Hypostereo.

Graphics processing unit (GPU): A high-performance 3D processor that integrates the entire 3D pipeline (transformation, lighting, setup, and rendering). A GPU offloads all 3D calculations from the CPU, freeing the CPU for other functions such as physics and artificial intelligence.

HDMI (high-definition multimedia interface): Similar to DVI (but using much smaller connectors), the multipin HDMI interface transfers uncompressed digital video with high-bandwidth digital copy protection and multichannel audio. Using an adapter, HDMI is backward-compatible with most current DVI connections, although any DVI-HDMI connection will pass video only, not audio [KIN200901].

Headset: A display device worn on the user's head, typically using LCD technology. These devices can be used in conjunction with a tracking device to create an immersive virtual reality.

Height error: *See* Vertical error.

HMD: Head-mounted display.

Holmes format: A format for stereo cards that are based on a stereoscope invented by Oliver Wendell Holmes. This is the format for most antique cards and has image centers that are further apart than the human eye (3-1/2″ × 7″). This is significant because any viewing device for such cards needs to have a mechanism for bending light before it reaches the eyes. Most viewers are prismatic.

Holmes stereoscope: Usual name for the common type of hand-held stereoscope with an open skeletal frame, named after its inventor in 1859, the American physician and author Oliver Wendell Holmes. Where, as is normally the case, the stereoscope includes a hood to shade the eyes and an adjustable card holder, it is more correctly termed a Holmes-Bates (or just Bates) stereoscope (after Joseph Bates, who introduced these refinements).

Holography: "Whole drawing." A technique for producing an image (hologram) that conveys a sense of depth but is not a stereogram in the usual sense of providing fixed binocular parallax information. Invented in theory by Dr. Dennis Gabor at Imperial College of London in 1948, holograms were not practical until the ruby laser was invented in 1960. Today, holograms are made with lasers and produce realistic images. Some appear to float in space in front of the frame, and they change perspective as one walks left and right. Holograms are monochromatic, and no special viewers or glasses are necessary, although proper lighting is important. To make a hologram, lengthy exposures are required with illumination by laser beams that must be carefully set up to travel a path with precisely positioned mirrors, beam splitters, lenses, and special film.

Holoscopic imaging: *See* Integral imaging.

Homologous points: Identical features in the left and right image points of a stereopair. The spacing between any two homologous points in a view is referred to as the separation of the two images (which varies according to the apparent distance of the points) and this can be used in determining the correct positioning of the images when mounting as a stereopair.

Homologues: *See* Homologous points.

Horizontal image translation (HIT): The horizontal shifting of the two image fields to change the value of the parallax of corresponding points. The term *convergence* has been confusingly used to denote this concept.

HUD: Head-up display; a display device that provides an image floating in midair in front of the user.

Hyperfocal distance: The distance setting on the focusing scale of a lens mount that will produce a sharply focused image from infinity to half the distance of the focus setting at any specific lens aperture. Of particular value in stereo photography to ensure maximum depth of field, so that viewing is not confused by out-of-focus subject matter.

Hyperstereo: Use of a longer than normal stereo base in order to achieve the effect of enhanced stereo depth and reduced scale of a scene; it produces an effect known as Lilliputism because of the miniaturization of the subject matter that appears as a result. Often used in order to reveal depth discrimination in architectural and geological features. The converse of hypostereo.

Hypostereo: Using a baseline that is less than the distance between the left and right eyes when taking the pictures. This exaggerates the size of the objects, making them look larger than life. It produces an effect known as giantism. The converse of hyperstereo. A good use for this would be 3D photographs of small objects; one could make a train set look life size.

Image splitter: A device mounted on the front of a single lens that, through the use of mirrors or prisms, divides the image captured on film into two halves, which are the two images of a stereoscopic pair. Sometimes called a frame splitter, and often imprecisely called a beam splitter.

Immersive: A term used to describe a system that is designed to envelop the participant in a virtual world or experience. The amount of immersion the participant feels depends on a number of factors. Visual immersion is the most common goal. This can be done effectively using a large screen or a head-mounted display.

Infinity, stereo: *See* Stereo infinity.

Integral imaging: A technique that provides autostereoscopic images with full parallax. This is a 3D photography technique described in 1908 by M. G. Lippmann. In the capture subsystem, an array of microlenses generates a collection of 2D elemental images onto a matrix image sensor (such as a CCD camera); in the reconstruction/display subsystem, the set of elemental images is displayed in front of a far-end microlens array, providing the viewer with a reconstructed 3D image. Also called integral photography.

Integral videography: A 3D imaging technique, integral photography is being extended to motion pictures.

Interaxial distance: Also called interaxial separation; the distance between camera lenses' axes.

Interlaced: A type of video stream made up of odd and even lines (or sometimes columns). Normal TV signals (such as PAL and NTSC) are interlaced signals, made up of two odd and even line images called fields. These odd and even fields can be used to store stereoscopic left and right images, a technique used on 3D DVDs, although this halves the vertical resolution of the video.

Interlens separation: The distance between the optical centers of the two lenses of a stereo camera or stereoscope, or (in wide-base stereography) between two photographic or viewing positions. Similar to base stereo.

Interocular adjustment: A provision in some stereo viewers that allows for adjustment of the distance between the lenses of the viewer to correspond with the image's infinity separation and in some cases the distance between a viewer's eyes.

Interocular distance: The separation between the optical centers of a twin-lens stereo viewer (which may be adjustable). Not necessarily the same as the interpupilary distance of the eyes.

Interocular separation: *See* Interpupilary distance.

Interpupilary distance (IPD): The distance between the centers of the pupils of the eyes when vision is at infinity. IPDs can range from 55 to 75 millimeters in adults, but the average is usually taken to be around 65 mm, the distance used for most resolving calculations and viewer designs.

Interpupilary separation: *See* Interpupilary distance.

Inversion: The visual effect achieved when the planes of depth in a stereograph are seen in reverse order; e.g., when the left-hand image is seen by the right eye, and vice versa. Often referred to as pseudostereo.

IR transmitter: A device that sends synchronization infrared (IR) signals to wireless shutter glasses.

JPEG: Joint Photographic Experts Group. An image format that reduces image size at the expense of throwing out information. The algorithm can be lossy or lossless. Most of the time, the loss of information is not noticeable. When saving an image, one can set the degree of compression one would like, at the expense of image quality. Usually, one can achieve 3:1 compression without noticing much degradation in resolution. JPEG uses an 8x8 grid and does a discrete cosine transformation on the image. When compression is high and quality is low one discerns a tiling pattern and visible artifacts at high-contrast boundaries, particularly noticeable in skies.

JPEG2000: A newer, more computationally intensive JPEG standard. It allows for higher compression rates than JPEG for comparable image quality loss. To achieve this, it uses a wavelet transformation on the image, which takes much more computing power, but as time progresses and machines become faster, this is less of a problem than when the first JPEG standard came out. The size of the compressible area can vary, so no tiling pattern is evident.

JPS: Stereoscopic JPEG file. A stereoscopic image file format that is based on JPEG compression.

Keystoning: Term used to describe the result arising when the film plane in a camera or projector is not parallel to the view or screen. The perspective distortion that follows from this produces an outline of, or border to, the picture, which is trapezoidal in shape and resembles the keystone of a masonry arch. In stereo, the term is applied to the taking or projecting of two images where the cameras or projectors are "toed-in" so that the principal objects coincide when viewed. The proportions of the scene will then have slight differences that produce some mismatching of the outlines or borders of the two images. Gross departures from orthostereoscopic practice (e.g., if using telephoto lenses) can produce keystoning in depth; more properly called a frustum effect.

LCD (liquid crystal display): Liquid crystal display technology is one of the methods used to create flat-panel TVs. Light is not created by the liquid crystals; a "backlight" behind the panel shines light through the display. The display consists of two polarizing transparent panels and a liquid crystal solution sandwiched in between. An electric current passed through the liquid causes the crystals to align so that light cannot pass through them. Each crystal acts like a shutter, either allowing light to pass through or blocking the light. The pattern of transparent and dark crystals forms the image [KIN200901].

LCoS (liquid crystal on silicon): A projection TV technology based on LCD. With LCoS, light is reflected from a mirror behind the LCD panel rather than passing through the panel. The control circuitry that switches the pixels on and off is embedded further down in the chip so it does not block the light, which improves brightness and contrast. This multilayered microdisplay design can be used in rear-projection TVs and projectors. TV makers use different names for their LCoS technologies—Sony uses SXRD, while JVC uses D-ILA and HD-ILA [KIN200901].

LDV (layered depth video): Layered depth video is a derivative and alternative to MV+D. It uses one color video with associated depth map and a background layer with associated depth map. The background layer includes image content that is covered by foreground objects in the main layer. LDV might be more efficient than MV+D because less data have to be transmitted. On the other hand additional error-prone vision tasks are included that operate on partially unreliable depth data, which may increase artifacts.

Lenticular: Pertaining to a lens; as used by Brewster to describe his lensed stereoscope; shaped like a lens. In stereo, used to describe: (1) a method of producing a depth effect without the use of viewing equipment, using an overlay of semi-cylindrical (or part-cylindrical) lens-type material that exactly matches alternating left and right images on a specially produced print, thereby enabling each eye to see only one image from any viewing position, as in an autostereogram, and (2) a projection screen with a surface made up of tiny silvered convex surfaces that spread the reflected polarized light to increase the viewing angle.

Lenticular screen: A projection screen that has embossed vertical lines for its finish rather than the "emery board" finish most common. They tend to cost more. The silvered version is critical to 3D projection, as any white screen will not preserve the polarization of the image reflected off it.

Lilliputism: Jargon term for the miniature model appearance resulting from using a wider-than-normal stereo base in hyperstereography.

Linear polarization: A form of polarized light in which the tip of the electric vector of the light ray remains confined to a plane.

Lorgnette: A handheld pair of lenses that helps people view stereographs.

Lumen: The unit of measure for light output of a projector. Different manufacturers may rate their projectors' light output differently. "Peak lumens" is measured by illuminating an area of about 10 percent of the screen size in the center of the display. This measurement ignores the reduction in brightness at the sides and corners of the screen. The more conservative "ANSI lumens" (American National Standards Institute) specification is made by dividing the screen into nine blocks, taking a reading in the center of each, and averaging the readings. This number is usually 20–25 percent lower than the peak lumens measurement [KIN200901].

Luminance: The brightness or black-and-white component of a color video signal; determines the level of picture detail [KIN200901].

Macro stereo: Ultra close-up images, photographed with a much-reduced stereo base in order to maintain correct stereo recession.

Macro stereo photography: Stereo photography in which the image on the film is about the same size or larger than the true size of the image.

Magic Eye: Paintings and computer generated optical illusions that, if one can free view, reveal hidden images of shapes and objects.

Mastering: Mastering (also known as content preparation) is the process of creating the master file package containing the movie's images, audio, subtitles, and metadata. Mastering standards are typically used in this process. Generally mastering is performed on behalf of the distributor/broadcaster at a mastering facility. Encryption may be added at this stage.

The file package is thus ready for duplication (e.g., DVDs) or for transmission via satellite or fiber links.

Mirror stereoscope: A stereo viewer incorporating angled mirrors, as in the Wheatstone and Cazes stereoscopes.

Misalignment: In stereo usage, a condition where one homologue or view is higher or lower than the other. Where the misalignment is rotational in both views, there is tilt; in one view only, twist. Viewing a misaligned stereogram can result in diplopia or produce eyestrain.

Monocular areas: Parts of the scene in a stereo image that appear in one view and not in the other. These can be natural (if behind the stereo window) or unnatural, as in the case of floating edges (if in front of the stereo window).

Monocular cues: *See* Extrastereoscopic cues.

Mount: In stereo usage, a special holder or card used to secure, locate, and protect the two images of a stereopair. Usually, the term includes any framing device or mask that may be incorporated.

Mounting: The process of fixing the left and right views to a mask or mount (single or double) so that they are in correct register, both vertically (to avoid misalignment) and horizontally (so that the stereo view is held in correct relationship to the stereo window).

Mounting jig: A device used to assist in the process of mounting stereopairs in correct register, usually incorporating an alignment grid placed below the mount holder and a pair of viewing lenses above the film chips to enable each eye to focus on the appropriate image and fuse the pairs.

MPEG: Standards developed by Moving Picture Experts Group. A type of audio/video file found on the Internet. There are three major MPEG standards: MPEG-1, MPEG-2, and MPEG-4.

Multiplex: The process of taking a right and left image and combining them with a multiplexing software tool or with a multiplexer to make one stereo 3D image.

Multiplexing: The technique for placing the two images required for a stereoscopic display within an existing bandwidth.

MV+D (multiview video plus depth): Advanced 3D video applications are wide-range multiview autostereoscopic displays and free-viewpoint video, where the user can choose his or her own viewpoint; requires a 3D video format that allows rendering a continuum of output views or a very large number of different output views at the decoder. Multiview video does not support a continuum and coding is increasingly inefficient for a large number of views. Video plus depth (V+D) supports only a very limited continuum around the available original view since view synthesis artifacts increase dramatically with the distance of the virtual viewpoint. Therefore, a MV+D representation is defined for advanced 3D video applications. MV+D involves a number of complex and error-prone processing steps. Depth has to be estimated for the N views at the sender, and N color with N depth videos have to be encoded and transmitted. At the receiver, the data have to be decoded and the virtual views have to be rendered. The multiview video coding (MVC) standard-developed MPEG supports this format and is capable of exploiting the correlation between the multiple views that are required to represent 3D video.

Near point: The feature in a stereo image that appears to be nearest to the viewer.

Near point of accommodation: The closest distance from the eyes that reading material can be read. This distance varies with age. It is often measured in each eye separately and both eyes together. The results are compared to one another.

Near-point stress: The term used when close work is causing the individual unacceptable stress. This is often seen when the relationship between accommodation and convergence is maintained only by excessive effort. The response to this is either a tendency to avoid close work (known as evasion) or alternatively to use progressively more and more effort. This is typified by a tendency to get closer and closer to the work and then to suffer slower work rates, headaches, and eye discomfort. Writing often becomes labored and difficult, showing a tight pencil grip and excessive pressure. Individuals may complain of blurred vision, print getting smaller, colored fringes around text that sometimes moves on the page, and possibly double vision. There is often a generalized ocular discomfort and there can be complaints of feeling "washed out" after prolonged concentration. Symptoms can vary from day to day.

Nimslo: The brand name, taken from the surnames of inventors Jerry Nims and Allen Lo, for a camera system intended primarily to produce lenticular autostereo prints, incorporating four lenses to record the same number of images (each of four perforations' width) on 35 mm film. The name is often used to identify the size of mask or mount developed to hold four-perforation-wide pairs of transparencies made with this camera and its derivatives.

Nimslo format: A stereo format that uses stereopairs of 4.5 perforations (film sprockets) per image width. This would be the equivalent of a half frame and is used with Nishika and Nimslo stereo cameras. Some cameras with beam splitters use a four-perforation format but this would not be called a Nimslo format.

NTSC: A type of interlaced video stream used primarily in North America. It is made up from 525 horizontal lines playing at 30 frames per second (or 60 fields per second).

Occlusion: Occlusion is the situation of blocking, as in when an object blocks light.

OLED (organic light emitting diode): OLED is an up-and-coming display technology that can be used to create flat-panel TVs. An OLED panel employs a series of organic thin films placed between two transparent electrodes. An electric current causes these films to produce a bright light. A thin-film transistor layer contains the circuitry to turn each individual pixel on and off to form an image. The organic process is called electrophosphorescence, which means the display is self-illuminating, requiring no backlight. OLED panels are thinner and lighter than current plasma or LCD HDTVs and have lower power consumption. Only small OLED screens are available at this time, but larger screens should be available by late 2010 [KIN200901].

One-in-30 rule: A rule-of-thumb calculation for determining the stereo base when using a nonstandard camera lens separation, e.g., in hyper- or macro stereography. To achieve optimum stereo depth, the separation of the centers of the camera lenses should be around 1/30 of the distance from the lenses to the closest subject matter in a scene. This "rule" only holds up under certain optical conditions (e.g., where "standard" focal-length lenses are used), and usually needs to be varied when, for example, lenses of longer or shorter than normal focal length are used.

OpenGL: A graphics application program interface (API) that was originally developed by Silicon Graphics, Inc. for use on professional graphics workstations. OpenGL subsequently grew to be the standard API for computer-aided design (CAD) and scientific applications, and today is popular for consumer applications such as PC games as well.

Orthoscopic image: A stereoscopic image viewed with its planes of depth in proper sequence, as opposed to an inverse (or pseudo) stereoscopic image.

Orthostereo: The ideal position and distance for viewing a stereo image.

Orthostereoscopic image: An image that appears to be correctly spaced as in the original view.

Orthostereoscopical viewing: When the focal length of the viewer's lenses is equal to that of the focal length of the taking lenses of the camera in which the slides were viewed. This is said to allow one to see the objects as being exactly the same size and with the same distance between each other in the viewer as in reality.

Over/under format: Over/under format involves using a mirror system to separate the left and right images that are placed one above one another. Special mirrored viewers are made for over/under format.

Over-and-under: A form of stereo recording (on cine film) or viewing (of prints) in which the left and right images are positioned one above the other rather than side by side, and viewed with the aid of prisms or mirrors that deflect the light path to each eye accordingly.

PAL: A type of interlaced video stream used in the United Kingdom and around the world. It is made up from 625 horizontal lines playing at 25 frames per second (or 50 fields per second).

Panum phenomenon: A trick of stereo viewing whereby, if a single vertical line is presented to one eye and two vertical lines to the other, and one of the double lines is fused with the single line in binocular viewing, the unmatched line is perceived to be nearer or further away than the fused line; a concept used in the design of stereo mounting grids; a phenomenon first described by the scientist Peter Ludvig Panum in 1858.

Parallax: Apparent change in the position of an object when viewed from different points; the distance between conjugate points. Generally, parallax is the set of differences in a scene when viewed from different points (as, photographically, between the viewfinder and the taking lens of a camera). In stereo, it is often used to describe the small relative displacements between homologues, more correctly termed deviation. When the left–right images in a pair lie directly atop one another, the object has zero parallax and appears on the plane of the screen. When the parallax is bigger than zero but smaller than the interocular separation, the object has positive parallax and appears to be behind the screen. A particular case happens when the parallax is equal to the interocular separation; this is called infinite parallax and objects appear to be at infinite distance. When the eyes' axis cross in front of the screen, negative parallax occurs. When parallax is greater than the interocular distance, the lines of sight diverge; this is called divergent parallax and never occurs in the real world [AUT200801].

Parallax budget: The range of parallax values, from maximum negative to maximum positive, that is within an acceptable range for comfortable viewing.

Parallax stereogram: A form of autostereogram that currently describes a technique in which alternate thin vertical strips of the left- and right-hand views are printed in a composite form and then overlaid with a grating (originally) or nowadays a lenticular sheet of cylindrical lenses that presents each view to the correct eye for viewing stereoscopically.

Parallel free-vision fusion: *See* Parallel method.

Parallel method: A free-viewing technique in which the lines of sight of the two eyes aim and meet at a point beyond and behind the 3D image; the eyes move outward (away from the nose) toward parallel lines of sight. It works with small images, but is somewhat limited on a computer screen.

Parallel viewing: *See* Parallel method.

Parallel viewing method: Viewing a stereo image where the left view of a stereo image is placed on the left and the right view is placed on the right. This is the way most stereocards are made as opposed to cross-eyed viewing.

Passive polarized 3D glasses: 3D glasses made with polarizing filters, used in conjunction with a view screen that preserves polarized light.

Passive stereo: A technique whereby 3D stereoscopic imagery is achieved by polarizing the left and right images differently at the source, viewed using low-cost polarizing glasses.

Photo bubble: *See* Photo cube.

Photo cube: A form of panorama picture made of photos usually taken with a fisheye lens. They are then stitched together to produce a photo sphere or cube. The viewer can see all around, above, and below.

Photo sphere: *See* Photo cube.

Photogrammetry: A discipline that uses stereography as a basis for scientific measurement and map-making; the art, science, and technology of obtaining reliable information about physical objects and the environment through processes of recording, measuring, and interpreting photographic images and patterns of recorded radiant electromagnetic energy and other phenomena.

Planar image: A planar image is one contained in a two-dimensional space but not necessarily one that appears flat. It may have all the depth cues except stereopsis.

Plano-stereoscopic: A stereoscopic projected image that is made up of two planar images.

Plasma: Plasma technology is one of the methods used to create flat-panel TVs. These are known as Plasma Display Panels (PDPs). The display consists of two transparent glass panels with a thin layer of pixels sandwiched in between. Each pixel is composed of three gas-filled cells or subpixels (one each for red, green and blue). A grid of tiny electrodes applies an electric current to the individual cells, causing the gas to ionize. This ionized gas (plasma) emits a high-frequency UV ray that stimulates the cells' phosphors, causing them to glow, which creates the TV image [KIN200901].

Polarization of light: The division of beams of light into separate planes or vectors by means of polarizing filters (first practically applied by Edwin Land of the Polaroid Company in the 1930s). When two vectors are crossed at right angles, vision or light rays are obscured.

Progression: With regard to film transport, progression is the amount or method by which film is advanced between exposures in a purpose-built stereo camera. The Colardeau progression moves by an even two frames; the Verascope progression moves by one and three frames alternately.

Projector: A video-display device that projects a large image onto a physically separate screen. The projector is typically placed on a table or is ceiling mounted. Projectors, sometimes referred to as front-projection systems, can display images up to 10 feet across, or larger. CRT-based projectors have generally been replaced by compact, lightweight, lower-cost digital projectors using DLP, LCD, or LCoS technology [KIN200901].

Pseudo: *See* Pseudoscopic.

Pseudo stereo: The effect produced when the left-view image and the right-view image are reversed. This condition causes a conflict between depth and perspective image.

Pseudoscopic: The presentation of three-dimensional images in inverse order, so that the farthest object is seen as closest and vice versa; more correctly referred to as inversion. Achieved (either accidentally or deliberately, for effect) when the left and right images are transposed for viewing.

Pseudoscopy: Viewing of stereopair images where the depth or relief of an object is reversed.

Pulfrich effect: Term now used to describe an illusory stereoscopic effect that is produced when two-dimensional images moving laterally on a single plane (as on a film or television screen) are viewed at slightly different time intervals by each eye, with the perceived delay between the eyes being achieved by means of reduced vision in one of them; e.g., through the use of a neutral-density filter. The apparent positional displacement that results from this is interpreted by the brain as a change in the distance of the fused image. A scene is produced giving a depth effect, the depth being proportionate to the rate of movement of the object, not to the object distance. The phenomenon was first adequately described in 1922 by Carl Pulfrich, a physicist employed by Carl Zeiss, Jena, in relation to a moving object (a laterally swinging pendulum).

Pulfrich stereo: Stereo video taken by rolling a camera sideways at a right angle to an object. When played back, the viewer wears glasses with one eye unobstructed, and the other through a darker lens. The brain is fooled into processing frames of the video in sequence, and the result is a moving stereo image in color.

R-mount: This is the name sometimes used to delineate the 41 × 101 mm, 1–5/8″ × 4″ (outer dimensions) mount used for almost all stereo slides. Mounts of these outer dimensions are made for the Realist, European, Nimslo, and full-frame formats. It is named after Seaton Rochwite, the inventor of the Realist stereo camera.

Ramsdell rig: *See* Beam splitter.

Random dot stereogram: A type of stereogram in which a three-dimensional image is formed by the fusing of apparently randomly placed dots in a stereopair, an effect first created manually by Herbert Mobbs of the Stereoscopic Society in the 1920s but scientifically developed, using computer-generated images, by Bela Julesz in the 1960s. The random dot stereogram is a computer-generated image that could be perceived *only* with binocular (two-eyed) depth perception. This is a method in which a pattern is repeated at about the distance between one's eyes (2.5–2.75 inches). Minor variations in the patterns from column to column will combine to give the viewer depth information when the eyes have diverged from their focus point (relaxed focus, walleyed). This method has limitations due to the fact that only graphics-type images can be shown, not a true-color image.

RBT (Raumbildtechnik GmbH): A leading manufacturer of stereo cameras, stereo projectors, and stereo systems.

Real-time 3D graphics: Real-time graphics are produced on-the-fly, by a 3D graphics card. Real time is essential if the user needs to interact with the images as in virtual reality, as opposed to watching a movie sequence.

Realist format: The five-perforation, 35 mm slide format of 23 × 24 mm, originally created by the specification of the Stereo Realist (United States) camera, and subsequently adopted by many other manufacturers. The stereo format uses stereopairs of five perforations (film sprockets) per image width. This is the most common stereo format and is named after the camera made by the David White Company. It is used with the Kodak, TDC Colorist I and II, TDC Vivid, Revere, Wollensak, Realist, along with many other cameras.

Rear projection: Rear projection occurs when images are projected from behind a screen. The advantage of this configuration is that a viewer cannot cast shadows by getting in between the projector and screen—particularly important when a user is interacting with images on the screen. Certain types of rigid and flexible rear-projection screens can be used for stereoscopic projection.

Rear-projection TV: Typically referred to as "big-screen" TVs, these large-cabinet TVs generally have built-in screens measuring at least 40 inches. Unlike the bulky CRT-based rear-projection TVs from years ago, today's "tabletop" rear-projection TVs are relatively slender and light. These TVs use digital microdisplay technologies like DLP, LCD, and LCoS [KIN200901].

Retinal disparity: *See* Disparity.

Retinal rivalry: Retinal rivalry is the simultaneous transmission of incompatible images from each eye.

Rig: Dual camera heads in a properly engineered mounting used to shoot stereo movies.

Rochwite mount: *See* R-mount.

Rotation: Tilting of the images by not holding the camera horizontally, causing one lens to be higher than the other at the picture-taking stage; a difference in the alignment of the two images in a stereogram caused by faulty mounting. If the tilting is not too severe, it may be possible to straighten both images when mounting but there will be a height error, however small, in part of the image.

Row interleaved: A format to create 3D video or images in which each row or line of video alternates between the left eye and the right eye (from top to bottom).

Savoy format: A stereo format produced by prisms or other forms of image splitter on a planar camera, side by side for still images and over-and-under for cine images.

Screen space: The region appearing to be within a screen or behind the surface of the screen. Images with positive parallax will appear to be in screen space. The boundary between screen and theater space is the plane of the screen and has zero parallax.

Selection device: The hardware used to present the appropriate image to the appropriate eye and to block the unwanted image. For 3D movies, the selection device is usually eyewear used in conjunction with a device at the projector, like a polarizing device.

Separation (interaxial): The distance between two taken positions in a stereo photograph; sometimes used to denote the distance between two homologues.

Septum: The partition used in a stereo camera to separate the two image paths; any partition or design element that effectively separates the lines of sight of the eyes such that only their respective left and right images are seen by each one.

Sequential stereograph: A stereopair of images made with one camera that is moved by an appropriate separation between the making of the left-hand and the right-hand exposures.

Shutter glasses: A device worn on the head, with two lenses generally covered in a liquid crystal material and controlled by a computer. When viewing a 3D image using these glasses, the computer displays the left image first, while instructing the glasses to open the left eye's "shutter" (making the liquid crystal transparent) and to close the right eye's "shutter" (making the liquid crystal opaque). Then in a short interval (1/30 or 1/60 of a second), the right image is displayed, and the glasses are instructed to reverse the shutters. This keeps up for as long as the viewer views the image. Since the time interval is so short, the brain cannot tell the difference in time and views them simultaneously. These glasses do not require polarized light-preserving screens.

Siamese: Used as a verb, to assemble a stereo camera from the relevant parts of two similar planar cameras; therefore, siamesed (adjective) to describe the finished assembly.

Silvered screen: A type of screen surface used for passive stereoscopic front projection. These screens maintain the polarization of the light introduced by polarizing filters in front of the two projector lenses.

Simulcast coding: The separate encoding (and transmission) of the two video scenes in the conventional stereo video (CSV) format.

Single-image random dot stereogram: A computer-generated stereogram in which the depth information is combined into a single image (a stereopair is no longer visible to the naked eye); a form of random dot stereogram in which the stereopair is encoded into a single composite image that each eye has to decipher separately. This was popularized in the "Magic Eye" types of books of the 1990s. The first single-image random dot stereogram was programmed on an Apple II computer in 1979 by Maureen Clarke and Christopher Tyler.

Slide bar: A device for taking sequential stereopairs of nonmoving subjects, enabling a planar camera to move by an appropriate separation while holding the camera in correct horizontal register with the optical axes either parallel or "toed-in" to create a convenient stereo window. It can be used to produce 2 × 2 stereo format slides.

Spinography: This is accomplished by walking around an object and taking pictures every 10 to 20 degrees, or putting the camera on a tripod and an object on a turntable and rotating it 10 to 20 degrees between shots. It can also be done with 3D modeling software by a computer, but does not create the same sense of depth as stereographics. To view spinography on a computer, one usually needs a small program for the browser called a plug-in.

Squeeze: Diminution of depth in a stereogram in relation to the other two dimensions, usually resulting from a viewing distance closer than the optimum (especially in projection). The opposite effect is to stretch.

Stereo: Having depth, or three-dimensions; used as a prefix to describe, or as a contraction to refer to, various stereographic or stereoscopic artifacts or phenomena. Stereo comes from the Greek *stereos* for hard, firm, or solid and it means *combining form, solid, three-dimensional*. Two inputs combine to create one unified perception of three-dimensional space.

Stereo acuity: The ability to distinguish different planes of depth, measured by the smallest angular differences of parallax that can be resolved binocularly.

Stereo blind: A term describing people who cannot fuse two images into one with depth (stereopsis).

Stereo infinity: The farthest distance at which spatial depth effects are normally discernible, usually regarded as 200 meters for practical purposes.

Stereo pair: In 1838 Charles Wheatstone invented the first stereoscopic viewer for the 3D viewing of stereo pairs.

Stereo vision: Two eye views combine in the brain to create the visual perception of one three-dimensional image; a by-product of good binocular vision. The separate images from two eyes are successfully combined into one three-dimensional image in the brain, also called stereoscopic vision. *See also* Stereopsis.

Stereo window: The viewing frame or border of a stereopair, defining a spatial plane through which the three-dimensional image can be seen beyond (or, for a special effect, "coming through"); a design feature in some stereo cameras whereby the axes of the lenses are offset slightly inward from the axes of the film apertures, so as to create a self-determining window in the resulting images that is usually set at around an apparent 2 m distance from the viewer. If the objects appear to be closer to the viewer than this plane, it is called breaking the window.

Stereocomparator: A stereoscopic instrument for measuring parallax; usually includes a means of measuring photographic coordinates of image points.

Stereogram: A general term for any arrangement of left-hand (LH) and right-hand (RH) views that produces a three-dimensional result, which may consist of: (1) a side-by-side or over-and-under pair of images, (2) superimposed images projected onto a screen, (3) a color-coded composite (anaglyph), (4) lenticular images, (5) a vectograph, and (6) in film or video, alternate projected LH and RH images that fuse by means of the persistence of vision.

Stereograph: The original term, coined by Wheatstone, for a three-dimensional image produced by drawing; now denoting any image viewed from a stereogram; in more general but erroneous usage as the equivalent of stereogram.

Stereographer: A person who makes stereo pictures.

Stereographoscope: An early type of stereoscope that also carries a large monocular lens (above the two regular stereoscopic lenses) for the viewing of planar photographs.

Stereography: The art and practice of three-dimensional image making.

Stereojet prints: Made of a special transparency material with polarized images ink-jetted onto each side, they can be displayed as transparencies or mounted against a reflective background and can be made up to poster size. They are viewed with an inexpensive pair of polarized lenses made for stereo viewing. Regular polarized sunglasses will usually not work because the lenses are mounted at the wrong angle of polarization. Colors are truer than anaglyphs, and when properly lit, they look three dimensional.

Stereopairs: In 1838 Charles Wheatstone invented the first stereoscopic viewer for the 3D viewing of stereopairs. Stereopairs are two images made from different points of view that are side by side. When viewed with a special viewer, the effect is remarkably similar to seeing the objects in reality.

Stereophotogrammetry: Stereophotogrammetry is based on the concept of stereo viewing, which derives from the fact that human beings naturally view their environment in three dimensions. Each eye sees a single scene from slightly different positions; the brain then "calculates" the difference and "reports" the third dimension.

Stereoplexing: A means to incorporate information for the left and right perspective views into a single information channel without expansion of the bandwidth.

Stereoplotter: An instrument for plotting a map or obtaining spatial solutions by observation of pairs of stereo photographs.

Stereopsis: The binocular depth sense, literally, "solid seeing." The blending of stereopairs by the brain. The physiological and mental process of converting the individual LH and RH images seen by the eyes into the sensation and awareness of depth in a single three-dimensional concept (or Cyclopean image).

Stereopticon: Term sometimes (erroneously) used to describe a stereoscope; first used (1875) to identify a dissolving twin-image magic lantern that could be used to convey information about depth by the blended sequential presentation of a series of planar views of a subject, and later applied to some other kinds of nonstereo projectors.

Stereorestitution: Process that uses two-dimensional information contained in a pair of images to re-create the shape and position of objects.

Stereoscope: A binocular optical instrument for helping an observer obtain the mental impression of a three-dimensional model when viewing plano-stereoscopic images (stereograms). The design of stereoscopic instruments uses a combination of lenses, mirrors, and prisms. It is usually an optical device with twin viewing systems.

Stereoscopic: "Solid seeing." Having visible depth as well as height and width; may refer to any experience or device that is associated with binocular depth perception; also called "solid looking."

Stereoscopic 3D: Two photographs taken from slightly different angles that appear three-dimensional when viewed together.

Stereoscopic multiplexing: *See* Stereoplexing.

Stereoscopy: The art and science of creating images with the depth sense stereopsis; the reproduction of the effects of binocular vision by photographic or other graphic means; stereography.

Strabismus: "Crossed eye," "walleye," "wandering eye," esotropia, exotropia, hyperphoria. It is a visual defect in which the two eyes point in different directions. One eye may turn either in, out, up, or down while the other eye aims straight ahead. Due to this condition, both eyes do not always aim simultaneously at the same object. This results in a partial or total loss of stereo vision and binocular depth perception. The eye turns may be visible at all times or may come and go. In some cases, the eye misalignments are not obvious to the untrained observer. Strabismus affects approximately 4 out of every 100 children in the United States.

Stretch: The elongation of depth in a stereogram in relation to the other two dimensions, usually caused by viewing from more than the optimum distance, especially in projection. The opposite effect is to squeeze.

Strip of stereo photographs: A series of overlapping photographs taken while moving the camera in one direction and at regular intervals so as to generate a sequence of stereo images.

Surround: The vertical and horizontal edges immediately adjacent to the screen.

t: In stereoscopy, *t* is used to denote the distance between the eyes, called the interpupilary or interocular distance. t_c is used to denote the distance between stereoscopic camera heads' lens axes and is called the interaxial.

Tautomorphic image: A stereoscopic image that presents the original scene to the viewer exactly as it would have been perceived in life; i.e., with the same apparent scale, positions of scenic elements, and a stereo magnification of x1 for all subject matter in the view.

Taxiphote viewer: A form of cabinet viewer devised by the Jules Richard Company for viewing a collection of stereograms in sequence and continuously.

Teco Nimslo: A camera that uses the Nimslo format but has been modified by Technical Enterprises to expose only two frames per exposure as opposed to the four frames per exposure needed for lenticular processing.

Theater space: The region appearing to be in front of the screen or out into the audience; can also be called audience space. Images with negative parallax will appear to be in theater space. The boundary between screen and theater space is the plane of the screen and has zero parallax.

Therapeutic 3D viewing: 3D viewing for the sake of improving important visual skills such eye teaming, binocular coordination, and depth perception.

Tissue: In stereo usage, an early type of stereogram on translucent paper in a card frame, often tinted and sometimes with pin-pricked highlights designed for viewing with backlighting.

Toeing-in: The technique of causing the optical axes of twin planar cameras to converge at a distance point equivalent to that of a desired stereo window, so that the borders of the images are coincident at that distance (apart from any keystoning that results).

Tracking: A 3D tracking system is used in virtual reality in order for the computer to track the participant's head and hands. There are many different types including optical, magnetic, and ultrasonic tracking systems.

Traditional photogrammetry: The use of film photography (usually diapositives) with analog or analytical stereoplotters.

Transcoding: The process of converting one 3D video format into another; for example, field sequential 3D video into column interleaved image data.

Transposition: The changing over of the inverted images produced by a stereo camera to the upright and left/right presentation necessary for normal viewing. This may be achieved optically by means of a transposing camera or viewer, or mechanically by means of a special printing frame, as well as manually during the mounting of images.

Tru-Vue: Proprietary name of a commercial stereo transparency viewing system that presents a series of views in a film-strip sequence on a single card mount.

Twin-camera stereo photography: Stereo photography using two monoscopic cameras, usually with shutters and other components connected internally or externally using mechanical or electronic means. This photography has advantages that include using common formats (e.g., full frame, medium format) and being able to achieve a variable stereo base. Drawbacks include difficulty matching cameras, film, and getting normal stereo bases. Camera bars can be used to help achieve more consistent results.

Twist: Rotational displacement of one view in a stereopair in relation to the other.

Two dimensional: *See* Planar image, 2D.

V+D (video plus depth): The video plus depth (V+D) representation consists of a video signal and a per-pixel depth map. (This is also called "2D-plus-depth" by some and "color plus depth" by others). Per-pixel depth data is usually generated from calibrated stereo or multiview video by depth estimation and can be regarded as a monochromatic, luminance-only video signal. The depth range is restricted to a range in between two extremes, Z_{near} and Z_{far}, indicating the minimum and maximum distance of the corresponding 3D point from the camera respectively. Typically, the depth range is quantized with 8 bits, associating the closest point with the value 255 and the most distant point with the value 0. With that, the depth map is specified as a gray-scale image, which can be fed into the luminance channel of a video signal and then be processed by any state-of-the-art video codec. For displaying V+D at the decoder, a stereopair can be rendered from the video and depth information by 3D warping with camera geometry information.

Vectograph: A form of polarization-coded stereogram (originally devised by the Polaroid Company) in which the images are mounted on the front and rear surfaces of a transparent base, and are viewed by polarized light or through polarized filters; the polarized equivalent of an anaglyph stereogram.

Verascope format: *See* Progression.

Vertical error: A fault present in a stereogram when the two film chips or prints are not aligned vertically in mounting, so that homologous points are at different heights.

Viewing angle: Measures a video display's maximum usable viewing range from the center of the screen, with 180 degrees being the theoretical maximum. Most often, the horizontal (side-to-side) viewing angle is listed, but sometimes both horizontal and vertical viewing angles are provided. For most home theater setups, horizontal viewing angle is more critical [KIN200901].

View-Master: Proprietary name of a commercial stereo transparency image display and viewing system utilizing stereopairs (seven in total) mounted in a circular rotating holder, and viewed with a purpose-made stereo viewer.

View-Master personal format: The format used with a View-Master Personal Camera. It produced two rows of chips of around 18 × 10 mm per roll of 35 mm film. These were used in conjunction with a cutter to make View-Master reels for personal use. It is not the same method that is used for mass-market reels produced by Fisher Price.

ViewMagic: Proprietary name of a commercial stereo print viewing system (utilizing angled periscope-type mirrors) for over-and-under mounted prints; the name now also being used to identify this mounting format.

Virtual reality: A system of computer-generated 3D images (still or moving) viewed by means of a headset linked to the computer that incorporates left-eye and right-eye electronic displays. The controlling software programs often enable the viewer to move interactively within the environment or "see" 360 degrees around a scene by turning the head, and also to "grasp" virtual objects in the scene by means of an electronically linked glove. Although they allow one to see all sides of an object by rotating it, one is still seeing only two dimensions at a time.

Vision: The act of perceiving and interpreting visual information with the eyes, mind, and body.

VRML: Virtual reality markup language. A set of standards for spinography software.

Wheatstone stereoscope: A "reflecting" or mirror stereoscope in which a pair of images (which need to be reversed) are placed facing each other at either end of a horizontal bar and viewed through a pair of angled mirrors fixed midway between them; named after Sir Charles Wheatstone, who devised this earliest form of stereoscope in 1832, prior to the advent of photography.

Widescreen: When used to describe a TV, widescreen generally refers to an aspect ratio of 16:9, which is the optimum ratio for viewing anamorphic DVDs and HDTV broadcasts [KIN200901].

Window: The stereo window corresponds to the screen surround unless floating windows are used.

Z-buffer: The area of the graphics memory used to store the Z or depth information about rendered objects. The Z-buffer value of a pixel is used to determine if it is behind or in front of another pixel. Z calculations prevent background objects from overwriting foreground objects in the frame buffer.

ZPS: Zero parallax setting or the means used to control screen parallax to place an object in the plane of the screen. ZPS may be controlled by horizontal image translation (HIT), or toe-in. One can refer to the plane of zero parallax, or the point of zero parallax (PZP) so achieved (prior terminology stated that left and right images are converged when in the plane of the screen).

References

[3DA201001] The 3D@Home Consortium. http://www.3dathome.org/

[AUT200801] Autodesk, Stereoscopic Filmmaking Whitepaper—The Business and Technology of Stereoscopic Filmmaking, 2008, Autodesk, Inc., 111 McInnis Parkway, San Rafael, CA 94903.

[COS200701] O.S. Cossairt, J. Napoli, S.L. Hill, R.K. Dorval, and G.E. Favalora, Occlusion-capable multiview volumetric three-dimensional display, *Appl. Optics*, 46, 8, 124ff, 10 March 2007.

[KIN200901] S. Kindig, TV and HDTV glossary, 2 December 2009, Crutchfield, Charlottesville, VA.

[MAT200901] K. Matsushima1 and S. Nakahara, Extremely high-definition full-parallax computer-generated hologram created by the polygon-based method, *Appl. Optics*, 48, 34, 1 December 2009.

Index

Milton Keynes UK
Ingram Content Group UK Ltd.
UKHW031143141024
449569UK00024B/1113